AUTOMOTIVE

TOOLS, FASTENERS, AND MEASUREMENTS

A TEXT-WORKBOOK

Books and Instructional Materials
by William H. Crouse and *Donald L. Anglin

Automotive Air Conditioning*
  Workbook for Automotive Air Conditioning*
Automotive Chassis and Body*
  Workbook for Automotive Chassis and Body*
Automotive Electrical Equipment
  Workbook for Automotive Electrical Equipment*
Automotive Emission Control*
  Workbook for Automotive Emission Control*
Automotive Engines*
  Workbook for Automotive Engines*
Automotive Fuel, Lubricating, and Cooling Systems*
  Workbook for Automotive Fuel, Lubricating, and Cooling Systems*
Automotive Tools, Fasteners, and Measurements—A Text-Workbook*
Automotive Transmissions and Power Trains*
  Workbook for Automotive Transmissions and Power Trains*
  Transparencies for Automatic-Transmissions Hydraulic Circuits*
Automotive Tuneup*
  Workbook for Automotive Tuneup*
Automotive Service Business: Operation and Management
Automotive Engine Design
Workbook for Automotive Service and Trouble Diagnosis
Automotive Mechanics
  Study Guide for Automotive Mechanics
  Testbook for Automotive Mechanics*
  Workbook for Automotive Mechanics*
  Automotive Engines—Sound Filmstrip Program
    Set 1 and Set 2
    Preview-Review Exercises
  Automotive Room Chart Series
    Automotive Electrical Equipment Charts
    Automotive Engines Charts
    Automotive Fuel Systems Charts
    Automotive Emissions Controls Charts
    Automotive Engines Cooling Systems, Heating, and Air Conditioning Charts
    Automotive Suspension, Steering, and Tires Charts
    Automotive Transmissions and Power Trains Charts
    Automotive Brakes Charts
  Automotive Troubleshooting Cards
The Auto Book
  Auto Shop Workbook*
  Auto Study Guide
  Auto Test Book*
  Auto Cassette Series
Pocket Automotive Dictionary*
General Power Mechanics (with Robert Worthington and Morton Margules*)
Small Engines: Operation and Maintenance
  Workbook for Small Engines: Operation and Maintenance

Automotive Transparencies by
William H. Crouse and Jay D. Helsel

Automotive Air Conditioning (In press)
Automotive Brakes
Automotive Electrical Systems
Automotive Emission Control (In press)
Automotive Engine Systems
Automotive Transmissions and Power Trains
Automotive Steering Systems
Automotive Suspension Systems
Engines and Fuel Systems

AUTOMOTIVE

TOOLS, FASTENERS, AND MEASUREMENTS

A TEXT-WORKBOOK

William H. Crouse

Donald L. Anglin

McGRAW-HILL BOOK COMPANY
Gregg Division
NEW YORK
ST. LOUIS
DALLAS
SAN FRANCISCO
AUCKLAND
BOGOTÁ
DÜSSELDORF
JOHANNESBURG
LONDON
MADRID
MEXICO
MONTREAL
NEW DELHI
PANAMA
PARIS
SÃO PAULO
SINGAPORE
SYDNEY
TOKYO
TORONTO

ABOUT THE AUTHORS

William H. Crouse

Behind William H. Crouse's clear technical writing is a background of sound mechanical engineering training as well as a variety of practical industrial experience. After finishing high school, he spent a year working in a tinplate mill. Summers, while still in school, he worked in General Motors plants, and for three years he worked in the Delco-Remy division shops. Later he became Director of Field Education in the Delco-Remy Division of General Motors Corporation for which he prepared service bulletins and educational literature.

During the war years, he wrote a number of technical manuals for the Armed Forces. After the war, he became Editor of Technical Education Books for the McGraw-Hill Book Company. He has contributed numerous articles to automotive and engineering magazines and has written many outstanding books about science and technology. He was the first Editor-in-Chief of the 15-volume McGraw-Hill Encyclopedia of Science and Technology. In addition, he has authored more than fifty technical books including *Automotive Mechanics*, which has sold over a million copies. His books have been widely translated and used in automotive mechanics training throughout the world.

William H. Crouse's outstanding work in the automotive field has earned for him membership in the Society of Automotive Engineers and in the American Society of Engineering Education.

Donald L. Anglin

Trained in the automotive and diesel service field, Donald L. Anglin has worked both as a mechanic and as a service manager. He has taught automotive courses in high school, trade schools, community colleges, and universities. He has also worked as curriculum supervisor and school administrator for an automotive trade school. Interested in all types of vehicle performance, he has served as a racing-car mechanic and as a consultant to truck fleets on maintenance problems.

Currently he serves as editorial assistant to William H. Crouse, visiting automotive instructors and service shops. Together they have coauthored magazine articles on automotive education and several books in the McGraw-Hill Automotive Technology Series.

Donald L. Anglin is a Certified General Automotive Mechanic and holds many other licenses and certificates in heavy duty truck mechanics, automotive education, service, and related areas. His work in the automotive service field has earned for him membership in the American Society of Mechanical Engineers and the Society of Automotive Engineers. In addition, he is an automotive instructor at Piedmont Virginia Community College, Charlottesville, Virginia.

Library of Congress Cataloging in Publication Data
Crouse, William Harry, [date]
Automotive tools, fasteners, and measurements.

1. Automobiles—Maintenance and repair—Problems, exercises, etc. 2. Tools—Problems, exercises, etc. 3. Measuring instruments—Problems, exercises, etc. I. Anglin, Donald L., joint author. II. Title.
TL152.C693 629.28'7'028 76-26169
ISBN 0-07-014630-6

AUTOMOTIVE TOOLS, FASTENERS, AND MEASUREMENTS: A TEXT-WORKBOOK

1 2 3 4 5 6 7 8 9 0 W C W C 7 8 3 2 1 0 9 8 7 6

The editors for this book were Ardelle Cleverdon and Susan L. Schwartz, the designer was Dennis G. Purdy, the art supervisor was George T. Resch, and the production supervisor was Rena Shindelman. It was set in Melior by York Graphic Services, Inc.
Printed and bound by Webcrafters, Inc.

CONTENTS

		Page	Date Completed	Grade
	Preface	ix		
	How to Use This Workbook	x		
Job 1.	Safety in the Shop	1	______	____

PART 1 MEASURING TOOLS

In this part, we discuss the importance of correct measurements. You will learn how to use the various measuring tools of the automotive technician. All the parts that go into an automobile have exact measurements (length, width, diameter, thickness). Many of the parts (especially in engines) have extremely close fits. Some dimensions must be accurate to within 1/1,000 of an inch [0.03 mm]. If the measurements are not accurate, the parts may not fit. A failure will occur.

The automotive technician compares measurements made on the automobile with specifications issued by the manufacturer. The difference between the two tells the technician what, if anything, must be done. For example, if a clearance is within specifications, then nothing need be done. But if the clearance is excessive, it indicates wear that requires service work.

So automotive technicians must know how to take different measurements accurately and quickly. They must also know how to interpret these measurements for the job they are doing. The jobs that follow will help you understand how to take measurements, and what the measurements mean.

The United States is beginning to convert to the metric system from its present U.S. Customary System of measurements. Thus, this part includes information about the metric system (millimeter, centimeter, meter, kilometer). It also shows how the metric system relates to the U.S. Customary System (inch, foot, yard, mile).

		Page	Date Completed	Grade
Job 2.	Rules and Calipers	7	______	____
3.	The Metric System	11	______	____
4.	Feeler Gauges	17	______	____
5.	Micrometers	21	______	____
6.	Dial Indicators	27	______	____
7.	Vernier Calipers	31	______	____
8.	Depth, Small Hole, and Telescope Gauges	35	______	____
9.	Measuring Clearances with Shim Stock and Plastigage	37	______	____

PART 2 FASTENING DEVICES AND METHODS

In this part, you will study the fastening devices used to hold the different parts of automobiles and engines together. Many parts are held together by nuts and bolts or screws. In addition, lockwashers of various types, keys, cotter pins, splines, and snap rings are used to lock or position various parts together. Some parts are pinned or riveted together; certain body parts are welded together. Also, brazing and soldering play a part in the assembling of the automobile. Most of these devices and procedures are discussed in this part. Most of the jobs require "hands-on" work. That is, you do the work yourself.

Job 10. Automotive Fasteners 43 ______ _____
11. Assembling Devices 47 ______ _____
12. Soldering 51 ______ _____

PART 3 HAND TOOLS

In this part, we discuss the common hand tools that an automotive technician uses. These include screwdrivers, hammers, wrenches, pliers, punches, socket wrenches, torque wrenches, and pullers. Cutting tools, such as files, chisels, and hacksaws, are covered in Part 4, even though they may also be considered hand tools. This division emphasizes that the cutting tools all have the common purpose of removing metal from the work. The tools in this section are the basic tools used in disassembly and reassembly work.

It is important that the trained automotive technician know how to use these tools. As a student, you should practice as much as possible with these tools, and become expert in their use. With practice, using them will become almost second nature. When that time comes, you will be well along the road to developing a marketable skill.

Job 13. Screwdrivers 55 ______ _____
14. Hammers 57 ______ _____
15. Pliers 59 ______ _____
16. Wrenches 63 ______ _____
17. Socket Wrenches 67 ______ _____
18. Torque Wrenches 71 ______ _____
19. Punches 75 ______ _____
20. Pullers 79 ______ _____

PART 4 CUTTING TOOLS

In this part, you will learn how to use various cutting tools required in the automotive shop. These include chisels, hacksaws, files, and tapes and dies. All these cutting tools have the same basic purpose; they remove metal from the piece being worked on. In a sense, all cutting tools perform their jobs similarly. Thus, a chisel has one cutting edge, a hacksaw blade has a hundred or so, and a file may have several hundred cutting edges. But all these cutting edges are alike in action. As a cutting edge is forced along the surface of a piece of metal, it cuts into the metal, removing shavings or chips of metal.

It is important that you practice using the cutting tools described in the jobsheets in this part. After some practice, you will find that these tools are not hard to use.

Job 21. Chisels 83 ______ _____
22. Hacksaws 87 ______ _____
23. Files 91 ______ _____
24. Taps 95 ______ _____
25. Thread Dies 99 ______ _____
26. Stud Removers 103 ______ _____
27. Tubing Cutter and Flaring Tool 107 ______ _____

PART 5 POWER TOOLS

A lot of shop tools and equipment are powered by electricity, compressed air, or hydraulic pressure. Such tools and equipment include car lifts, grinders, hoists, tire changers, and grease guns. In this part we discuss the most frequently used shop power tools—impact wrenches, drills, and grinders.

Here are a few cautions to remember: The shop is equipped with a number of power tools. You should carefully follow the instructions for using each of these tools. Keep your hands and clothes away from moving machinery, such as engine fans, drive belts, and grinding wheels. Keep your hands out of the way when using cutting tools, such as cylinder-boring equipment and the drum lathe. Never attempt to feel the finish while the machine is running. There may be slivers of metal on the surface of the work, and they could cut your hand seriously. When working on equipment with compressed springs, such as clutches or valves, use care to keep the springs from slipping and jumping loose. They could fly off at high speed and hurt someone. And, never oil or adjust moving machinery unless the instructions specifically tell you to do so.

Job 28. Air Tools and Impact Wrenches 111 ______ ______

29. Electric Drills and Twist Drills 115 ______ ______

30. Bench Grinders 121 ______ ______

PART 6 METRIC MEASUREMENT

Metric measurement is coming into wider use in the United States. Throughout this workbook we have tried to increase your familiarity with metric measurements. As you have seen, each measurement in the U.S. Customary System is followed by its metric equivalent.

The purpose of this part is to give you practice in converting to metric measurements and using conversion charts. Also, it will help you to become familiar with the metric measurements most frequently used in the automotive shop.

31. Using the Metric System 127 ______ ______

32. Metric Engine Measurements 129 ______ ______

Metric Conversion Tables 131

Answer Key 134

PREFACE

Automotive Tools, Fasteners, and Measurements is a combination shop guide, study guide, notebook, and work-progress report. It is designed to cover the basic jobs on the subjects. It contains shop exercises in the selection, use, and care of the various hand and power tools employed by automotive technicians. By the time students have completed all the jobs detailed in this text-workbook, they should have some proficiency in the use of these tools. Ability to select and use hand tools is a basic requirement of a good automotive technician. It is the purpose of this text-workbook to help students gain that ability.

At the top of every Jobsheet is a student performance objective for use by the student and the instructor. By stating the objective for each Jobsheet, the student knows the task at hand. Then the instructor can measure how well the task was done. Each student can keep a personal progress chart by posting the date and grade of each completed Jobsheet in the space provided to the right of its title in the Table of Contents.

The safety cautions are not repeated for every job because doing so would take up undue space. However, the student must keep in mind all the safety cautions outlined in previous jobs while progressing through later ones. It is particularly important to wear goggles on any job where there is danger from flying particles, as when grinding or chipping. A single flying particle can cause blindness. No one should ever take chances. Goggles should always be worn whenever there is danger from flying particles. In addition, the locations of fire extinguishers should be called to students' attention, and the instructor should explain how to use them.

There are several types of questions in the text-workbook. Many are of the multiple-choice type. For the True-or-False Tests, the student should place a "T" in the blank if the statement is true and an "F" if it is false. Other questions require longer written answers. Answers to the Multiple-Choice and True-or-False Tests are at the end of the book.

The authors would like to thank the many people, both in education and in industry, who by their assistance and counsel have contributed so much to this text-workbook. It is hoped that *Automotive Tools, Fasteners, and Measurements* will help to achieve the aim of all who work in the field of automotive mechanics instruction: to train high-caliber automotive technicians capable of taking their proper place in the automotive servicing profession.

William H. Crouse
Donald L. Anglin

HOW TO USE THIS WORKBOOK

_____**1** As you read through the Jobsheet, notice that there is a line to the left of each numbered step in the procedure. Place a check mark above the line as you complete each step.

_____**2** Before attempting a job, read carefully any additional reference text that your instructor assigns. Read it more than once, if necessary, to gain a good understanding of the principles involved.

_____**3** Read the Jobsheet through to the questions. This will tell you the purpose of the job and how it is to be done.

_____**4** Make sure the equipment, tools, and supplies needed to complete the job are on hand. Check with the instructor if you are not sure that you have everything required.

_____**5** Note carefully the safety cautions. If references are made to safety cautions in previous jobs, be sure to reread those safety cautions. Do not assume that you remember them.

_____**6** Follow the procedure, step by step, as detailed in the Jobsheet. If anything is not clear, ask the instructor about it.

_____**7** After completing the job, answer the questions assigned by the instructor. When answering a multiple-choice question, select the word or phrase after the sentence that will correctly complete the sentence. Then put the letter that precedes the word or phrase into the blank space provided. For example:

1 Before the clutch is removed from the car, it is first necessary to remove the **A** engine **B** differential **C** transmission **D** rear wheels

1_____

Since "**C** transmission" is the correct answer, you would put a "C" in the blank space provided.

_____**8** When writing down the answers to a checkup question that asks you to explain a service procedure, do not copy from the Jobsheet or from the reference text. Instead, write down exactly what you did in performing the job. What you did may be somewhat different from the procedure given in the workbook or text because of some difference in the equipment or because of some special problem which came up during the job.

_____**9** Notice that there is room in the Table of Contents to write in the dates you complete the various jobs and the grades you are assigned by your instructor. This information will serve as your progress guide, which will tell you how you are getting along with your shop work.

William H. Crouse
Donald L. Anglin

NAME DATE CLASS

Objective: To learn about the automotive shop, and how to work safely in the shop.

Job 1

SAFETY IN THE SHOP

PROCEDURE

_____ **1** Shopwork is varied and interesting. In the shop, you will learn many vehicle and engine jobs. But before you work in the shop, you should know about safety. Safety in the shop means protecting yourself and others from possible danger and injury. In this jobsheet, we discuss some important safety rules that you should follow in the shop, to protect yourself from harm. Remember this: When everybody obeys the safety rules, the shop is a much safer place to work than your home! Many more people are hurt in the home than in the shop.

_____ **2** Safety is your job. In the shop, you are "safe" when you protect your eyes, your fingers, your hands—all of yourself—from danger. And, just as important, when you look out for the safety of those around you. Some very imporant safety rules are listed and discussed in the next few pages. Your instructor will discuss these and many more safety rules with you. Follow the rules for your own protection, and for the protection of your fellow workers.

_____ **3** There is one rule that you should always keep in mind in the shop. It is extremely important when you are working with tools or around vehicles. It is called the "ABC of Safety" (Fig. 1-1): Always Be Careful!

_____ **4** When you first go into the shop, take a good look around. Notice the shop layout. The term "shop layout" means the locations of workbenches, car lifts, power tools, and so on. Shop layouts vary (Fig. 1-2). So the first thing you should do in a shop is find out where everything is located. This includes the different power tools and the workbenches, car lifts, and work areas. Many shops have painted lines on the floor to mark off work areas. These lines guide customers and workers away from danger zones where machines are being operated. The lines also remind workers to keep their tools and equipment within work areas.

_____ **5** Many shops have warning signs posted around machinery. These signs are there to remind you about safety, and about how to use machines safely. Follow the posted instructions at all times. The most common cause of accidents in the shop is failure to follow instructions.

_____ **6** Locate the first-aid kits so that you can get one when it is needed. Note the location of the telephone and the nurse's office. This will allow you to get help quickly in case of an accident or fire. If there is an accident and someone gets hurt, notify your instructor at once. Your instructor knows what to do. Be very careful in attempting first aid. You must know what you are doing. If first aid is done wrong, it can do more harm than good. For example, a serious back injury could be made worse if the injured person is moved. On the other hand, quick mouth-to-mouth resuscitation may save the life of a person who has suffered an electric shock. Talk to your instructor if you have any questions about safety and first aid.

_____ **7** Notice the location of the fire extinguish-

Fig. 1-1. The ABC of safety in the shop.

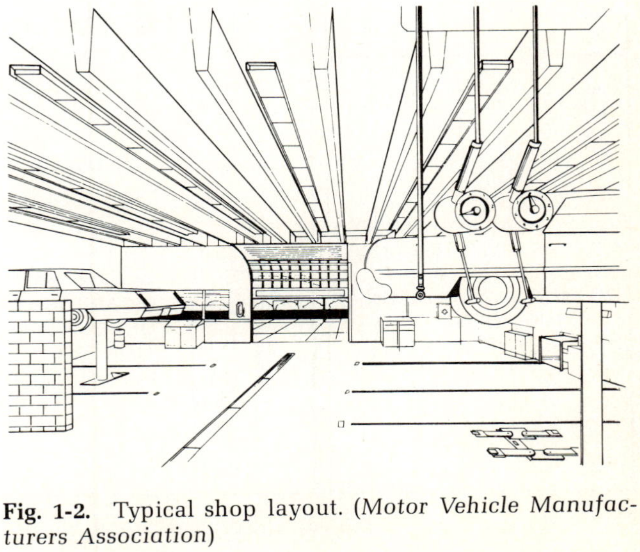

Fig. 1-2. Typical shop layout. (*Motor Vehicle Manufacturers Association*)

ers. Make sure you know how to use them. Figure 1-3 is a chart showing the different types of fires, and the types of fire extinguishers used to fight them. Take a moment right now to study Fig. 1-3. Remember this about fires: The quicker you get at them, the easier it is to control them. But you have to use the right kind of fire extinguisher, and use it correctly. Ask your instructor if you have any questions about fire extinguishers, where they are located, and how to use them.

_____ **8** Gasoline is used so much in the shop that people forget it is very dangerous if not handled properly. A spark or lighted match in a closed place filled with gasoline vapor can cause an explosion. Even the spark from a light switch can set off an explosion (Fig. 1-4). So you must always be careful with gasoline.

_____ **9** Suppose there is gasoline vapor around, because someone spilled gasoline or a fuel line is leaking. Then you should keep the shop doors open

FIRES

A	CLASS *A* FIRES ORDINARY COMBUSTIBLE MATERIALS SUCH AS WOOD, PAPER, TEXTILES AND SO FORTH. REQUIRES. . .COOLING–QUENCHING
B	CLASS *B* FIRES FLAMMABLE LIQUIDS, GREASES, GASOLINE, OILS, PAINTS AND SO FORTH. REQUIRES. . .BLANKETING OR SMOTHERING
C	CLASS *C* FIRES ELECTRICAL EQUIPMENT, MOTORS, SWITCHES AND SO FORTH. REQUIRES. . .A NONCONDUCTING AGENT

EXTINGUISHERS

TYPE	USE	OPERATION
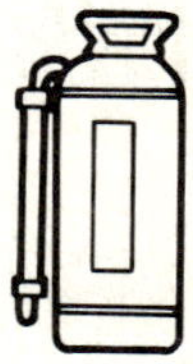 **FOAM** SOLUTION OF ALUMINUM SULPHATE AND BICARBONATE OF SODA	OK FOR **A B** NOT FOR **C**	*FOAM*: DON'T PLAY STREAM INTO THE BURNING LIQUID. ALLOW FOAM TO FALL LIGHTLY ON FIRE 
 **CARBON DIOXIDE** CARBON DIOXIDE GAS UNDER PRESSURE	NOT FOR **A** OK FOR **B C**	*CARBON DIOXIDE*: DIRECT DISCHARGE AS CLOSE TO FIRE AS POSSIBLE. FIRST AT EDGE OF FLAMES AND GRADUALLY FORWARD AND UPWARD 
 **DRY CHEMICAL**	MULTI-PURPOSE TYPE: OK FOR **A B C** ORDINARY BC TYPE: NOT FOR **A**; OK FOR **B C**	*DRY CHEMICAL*: DIRECT STREAM AT BASE OF FLAMES. USE RAPID LEFT–TO–RIGHT MOTION TOWARD FLAMES 
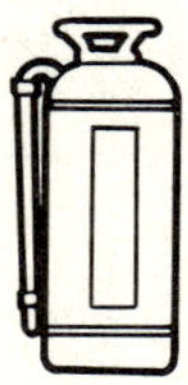 **SODA–ACID** BICARBONATE OF SODA SOLUTION AND SULPHURIC ACID	OK FOR **A** NOT FOR **B C**	*SODA–ACID*: DIRECT STREAM AT BASE OF FLAME 

Fig. 1-3. Chart showing types of fire extinguishers and the classification of fires. (*Ford Motor Company*)

Fig. 1-4. An explosion can result from gasoline fumes.

and keep the ventilating system running. Wipe up the spilled gasoline at once, and put the rags outside to dry. Never smoke or light cigarettes around gasoline. When you work on a leaky fuel line, carburetor, or fuel pump, catch leaking gasoline in a container or with rags. Put the soaked rags outside to dry. Fix the leak as quickly as possible. And don't make sparks around the car, for instance by connecting a 12-volt trouble light to the battery.

_____ **10** Store gasoline in an approved safety container (Fig. 1-5). Never, *never* store gasoline in a glass jug. The jug could be broken and could cause a terrible explosion or fire (Fig. 1-6).

_____ **11** Oily rags can also start fires. They can catch fire without a spark or flame. Oily rags and waste should be put into special closed metal containers. In these special containers, they can do no harm if they ignite from spontaneous combustion or another cause (Fig. 1-7).

_____ **12** Never siphon gasoline from a tank using your mouth and a piece of hose. Swallowing even a small amount of gasoline can cause respiratory infection and pneumonia. In addition, the lead added to gasoline is poisonous. If you should get some gasoline in your mouth, spit it out; rinse your mouth several times, and avoid taking deep breaths. If you swallow some gasoline, do not try to vomit. Get immediate help. Your life may depend on it!

Fig. 1-6. Glass bottles, jars, and jugs must never be used to store gasoline or other flammable liquids.

_____ **13** Some people say, "Accidents will happen!" But safety experts do not agree. They say, "Accidents are caused: They are caused by careless actions, by inattention to the job at hand, by using damaged or incorrect tools. And sometimes accidents are caused by plain stupidity!" Here are some simple safety rules that will help prevent accidents in the shop.

_____ **14** Work quietly, and give the job your full attention. This means no horseplay! Never indulge in horseplay or other foolish antics in the shop. You could cause yourself or someone else to get seriously hurt. Give your attention to the job you're doing.

_____ **15** Keep your tools and equipment under control. Jack handles should be kept up, out of the way. Don't leave tools, creepers, and other items on the floor where someone could trip on them and get hurt. After using a creeper, prop it against a wall or shove it under the car being worked on (Fig. 1-8).

_____ **16** Don't put sharp objects, such as screwdrivers, in your pocket. You could cut yourself or get stabbed. Or you could ruin the upholstery in a car.

_____ **17** Make sure your clothes are right for the

Fig. 1-5. Store gasoline and all flammable liquids in approved safety containers.

Fig. 1-7. Safety can for the storage of oily rags.

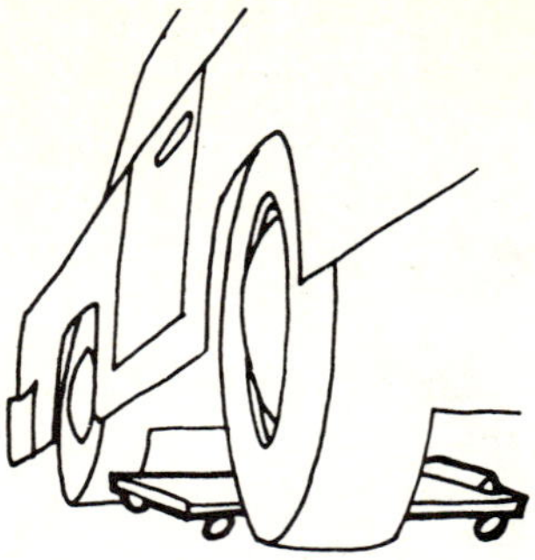

Fig. 1-8. After using a creeper, shove it under the car being worked on. (*Ford Motor Company*)

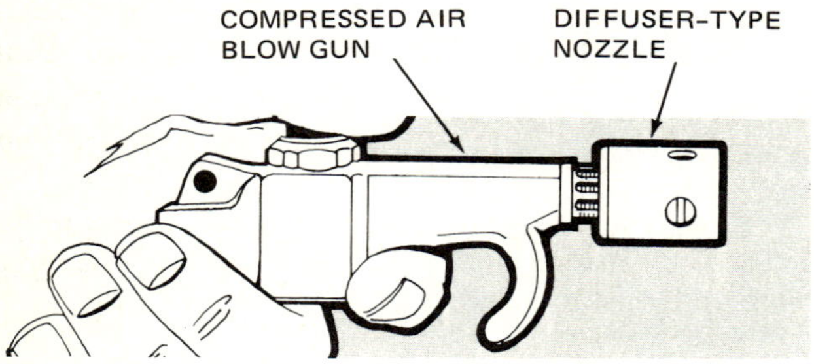

Fig. 1-9. Approved type of diffuser nozzle on a compressed-air blowgun. The maximum allowable discharge pressure is 30 psi [2.1 kg/cm^2]. (*Ford Motor Company*)

job. Dangling sleeves or ties can get caught in machinery and cause serious injuries. Keep long hair out of machinery by wearing a cap.

_____ **18** No bare feet in the shop! There are too many ways to get hurt in the shop without shoes. Do not wear sandals or open-toe shoes. Wear full leather shoes with nonskid rubber heels and soles. Steel-toe safety shoes are best for shop work.

_____ **19** Keep your hands and tools free of excess grease and oil. Otherwise, you will not be able to get a good grip on your tools.

_____ **20** If you spill oil, grease, or any liquid on the floor, clean it up so that no one will slip and fall.

_____ **21** Compressed air can be dangerous. Never use compressed air to blow dirt from your clothes. Never point the compressed-air blowgun at another person. Flying particles could put out an eye. The Occupational Safety and Health Act (OSHA) requires that compressed-air blowguns discharge air at a pressure of not more than 30 psi (pounds per square inch) [2.2 kg/cm^2 (kilograms per square centimeter)] (Fig. 1-9).

_____ **22** Always wear a face shield, safety glasses, or safety goggles when you are doing any job that could endanger your eyes (Fig. 1-10). Always wear suitable eye protection when using a grinding wheel, or when cutting or welding. Watch out for sparks from a grinding wheel or welding equipment. The sparks can set your clothes on fire.

_____ **23** To protect your eyes, wear safety goggles when using any chemical cleaner, such as solvent or carburetor cleaner. If you get a chemical in your eyes, wash them with water at once (Fig. 1-11). Then see the school nurse or a doctor as soon as possible.

_____ **24** When using a jack to lift a car, make sure the jack is properly positioned. Make sure it won't slip and let the car fall. And never, *never* jack up a car while someone is working under it! People have been killed when the jack slipped and the car fell on them! Always place car stands or supports in proper position under the car, before you go under it yourself.

_____ **25** Never run an engine in a closed garage unless a working shop ventilation system is hooked up to the tail pipe of the car. Engine exhaust gases contain carbon monoxide (CO). Carbon monoxide is a colorless, tasteless, odorless, poisonous gas that can kill you! In a closed one-car garage, enough carbon monoxide to kill you can collect in only 3 minutes.

_____ **26** Remember your personal responsibility to work safely while you are in the shop. Look out for yourself and for the safety of those working around you. In other words, Always Be Careful.

Fig. 1-10. Always wear safety glasses, goggles, or a face shield when using a bench grinder. (*Ford Motor Company*)

Fig. 1-11. If solvent or some other chemical splashes in your eye, immediately wash your eye with water.

Multiple-Choice Test — Job 1

1 Horseplay in the shop is **A** permitted **B** seldom permitted **C** never permitted **D** frequently permitted **1** ______

2 Carrying screwdrivers in your back pockets can **A** injure you **B** keep tools handy **C** keep tools clean **D** protect the tools **2** ______

3 Long, loose sleeves worn around moving machinery could **A** keep you clean **B** improve your appearance **C** get caught in the machinery **D** protect the machine **3** ______

4 Tools may slip out of your hands if your hands are covered with **A** gloves **B** sand **C** grease or oil **D** protective cream **4** ______

5 A compressed-air nozzle blowgun must never be **A** used in the automotive shop **B** pointed at someone else **C** left pressurized overnight **D** kept on a reel **5** ______

6 The right time to jack up a car is when **A** someone is working under it **B** never **C** several people are working under it **D** no one is working under it **6** ______

7 Before you get under a car to work on it, you must make sure that **A** the jack handle is up **B** you use a creeper **C** the tires are inflated **D** the car is supported on stands **7** ______

8 Creepers left sticking out from under a car can **A** get dirty **B** keep the car from rolling **C** cause someone to trip **D** add speed to your work **8** ______

9 You should wear safety goggles **A** when doing any job that could endanger your eyes **B** when told to wear them by the instructor **C** only if you do not wear glasses **D** while welding **9** ______

10 Oil-soaked rags should be **A** washed in the sink **B** stored in a special safety container **C** thrown in a corner **D** used to mop the floor **10** ______

11 Bare feet are not allowed in the automotive shop because of **A** the danger of injury **B** untrimmed toenails **C** the school dress code **D** athlete's foot **11** ______

12 An engine running in a closed room is dangerous because the exhaust gases contain **A** blue smoke **B** water **C** air **D** carbon monoxide **12** ______

13 In shop safety, ABC stands for the words **A** Always Buy Cotton **B** Always Be Careful **C** Alcoholic Beverage Control **D** Auto Buyers Club **13** ______

14 When working in the automotive shop, your attention should be directed to **A** the clock **B** your friends **C** the job you are doing **D** the car parked outside **14** ______

15 Fewer accidents occur in a shop that is **A** cluttered **B** neat and clean **C** overcrowded **D** poorly lighted **15** ______

Written Checkup — Job 1

1 Write down the location of each fire extinguisher in the shop.
2 Describe how to use the types of fire extinguishers in your shop.
3 Locate the nearest fire alarm. Write down its location.
4 Write down the location of each first-aid kit in the shop. Describe what type of injuries can be treated with a first-aid kit.
5 Whom do you notify when you think someone has been seriously injured in the shop? How do you notify them?

NAME DATE CLASS

Objective: To learn how to use rules and calipers.

Equipment: Round stock or shafts, rectangular stock, plates or machine parts with holes in them.

Tools: Steel rules, inside and outside calipers.

Job 2

RULES AND CALIPERS

Part One MEASUREMENTS

_____ **1** Almost any service job performed by a mechanic or a technician requires the taking of measurements. In some cases, thickness, diameter, or length must be measured. In other cases, engine power output, vacuum, or speed has to be measured. Taking measurements is often the first step in any testing procedure or repair job.

_____ **2** Many measurements are made in *inches* or fractions of an inch. However, as the United States begins to use the metric system, the *meter* and its divisions will become familiar to the mechanic. We shall discuss the metric system of measurement in Job 3. Now, let's discuss rules and calipers, two frequently used measuring tools.

Part Two RULES

_____ **1** The common rule, or steel scale (Fig. 2-1), is marked off in inches and fractions of an inch. Sometimes these markings are as small as $\frac{1}{64}$ in (inch). Other rules or steel scales are marked in both inches and centimeters (or millimeters). The metric rule has only metric markings (Job 3).

_____ **2** The rule is not of great use to the mechanic or technician. It has two serious disadvantages. One, it is not accurate enough to measure many machine parts. Two, it cannot be used to take certain measurements, because of its shape. However, the 6-in steel scale is very useful in shop work.

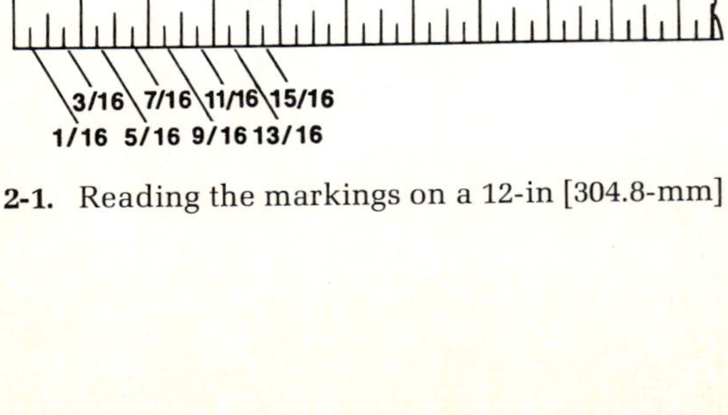

Fig. 2-1. Reading the markings on a 12-in [304.8-mm] rule.

_____ **3** Examine a steel rule (also called a scale). Note the inch and fractions-of-an-inch markings (Fig. 2-1).

_____ **4** Note that the longest line between the inch marks is the $\frac{1}{2}$-in mark. Note that shorter lines are used for the $\frac{1}{4}$-in marks. Still shorter lines are used for the $\frac{1}{8}$-, $\frac{1}{16}$-, $\frac{1}{32}$-, $\frac{1}{64}$-inch marks (Fig. 2-1).

_____ **5** Some rules have more than one scale. For instance, many rules have one edge that is divided up into no more than eighths of an inch. A second edge is divided into sixteenths of an inch; a third edge into thirty-seconds of an inch; and a fourth edge into sixty-fourths of an inch. Some uses of the rule are shown in Fig. 2-2.

_____ **6** Measure various pieces, as shown, and record the measurements below.

MEASUREMENTS WITH A STEEL RULE	
Piece 1	_____ in
Piece 2	_____ in
Piece 3	_____ in
Piece 4	_____ in
Piece 5	_____ in

_____ **7** Another type of rule has one edge divided into decimal fractions of an inch. That is, each inch is divided into fifty equal parts. This allows you to read a measurement of 0.020 in (twenty thousandths of an inch) directly from the rule. The decimal-equivalent chart (Table 2-1) shows that this is slightly more than $\frac{1}{64}$ in.

Part Three CALIPERS

_____ **1** Calipers can be used (with a scale) to make fairly accurate measurements of many parts. The outside caliper (Fig. 2-3) can be used to measure the diameter of a shaft.

_____ **2** First, it is adjusted by turning the thumb nut until the legs just slip over the shaft with a slight drag (Fig. 2-3). Then, the distance between the two legs is measured with a scale (Fig. 2-4).

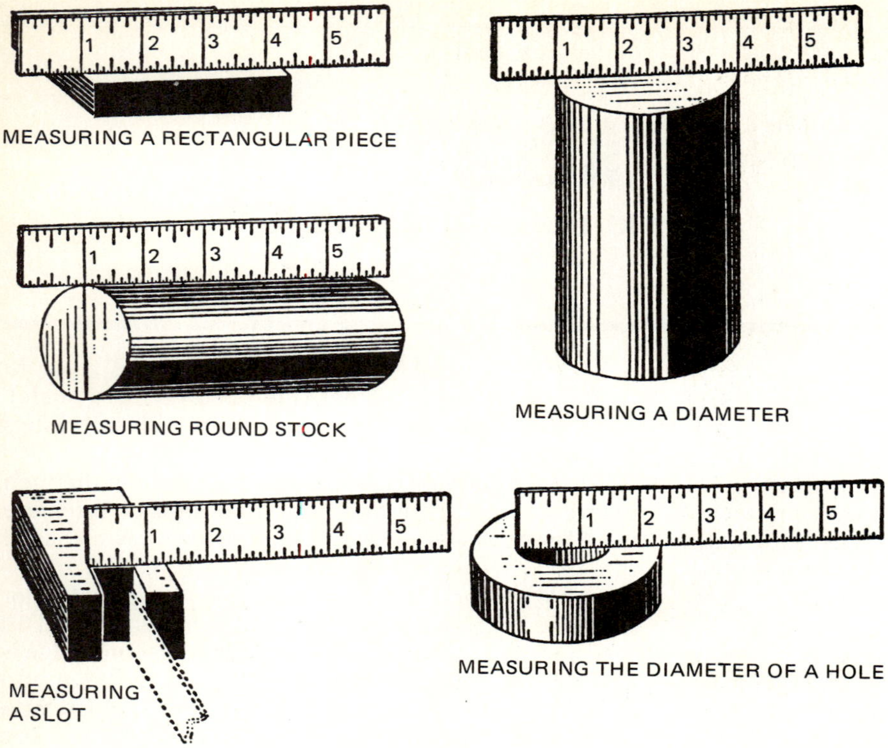

Fig. 2-2. Various uses of a 6-in [152.4-mm] rule.

NOTE: Do not tighten the calipers so much that they must be forced over the round stock. This springs the calipers and prevents accurate measurements.

_____ **3** Measure various pieces, as shown, and record the measurements below.

MEASUREMENTS WITH OUTSIDE CALIPER	
Piece 1	_____ in
Piece 2	_____ in
Piece 3	_____ in
Piece 4	_____ in
Piece 5	_____ in

_____ **4** In a similar manner, the inside caliper can be used to measure the diameter of a hole (Fig. 2-5). It is adjusted until the legs can be moved in the hole with a slight drag. Then, the distance between the legs is measured with a scale (Fig. 2-6). Note that the scale used in Fig. 2-6 reads in decimal fractions of an inch.

_____ **5** Measure various pieces, as shown, and record the measurements in the table on p. 9. Remember, when using the inside caliper, to read the measurement from the steel rule.

TABLE 2-1 DECIMAL EQUIVALENTS

1/64 0.0156	17/64 0.2656	33/64 0.5156	49/64 0.7656
1/32 0.0312	9/32 0.2812	17/32 0.5312	25/32 0.7812
3/64 0.0468	19/64 0.2969	35/64 0.5469	51/64 0.7969
1/16 0.0625	5/16 0.3125	9/16 0.5625	13/16 0.8125
5/64 0.0781	21/64 0.3281	37/64 0.5781	53/64 0.8281
3/32 0.0937	11/32 0.3437	19/32 0.5937	27/32 0.8437
7/64 0.1094	23/64 0.3594	39/64 0.6094	55/64 0.8594
1/8 0.125	3/8 0.375	5/8 0.625	7/8 0.875
9/64 0.1406	25/64 0.3906	41/64 0.6406	57/64 0.8906
5/32 0.1562	13/32 0.4062	21/32 0.6562	29/32 0.9062
11/64 0.1719	27/64 0.4219	43/64 0.6719	59/64 0.9219
3/16 0.1875	7/16 0.4375	11/16 0.6875	15/16 0.9375
13/64 0.2031	29/64 0.4531	45/64 0.7031	61/64 0.9531
7/32 0.2187	15/32 0.4687	23/32 0.7187	31/32 0.9687
15/64 0.2344	31/64 0.4844	47/64 0.7344	63/64 0.9843
1/4 0.25	1/2 0.5	3/4 0.75	1 1.0

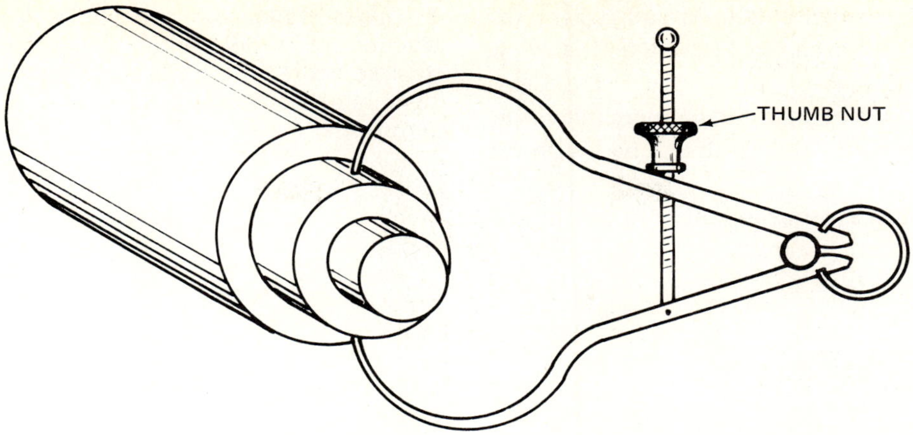

Fig. 2-3. Using outside calipers to measure the diameter of a shaft.

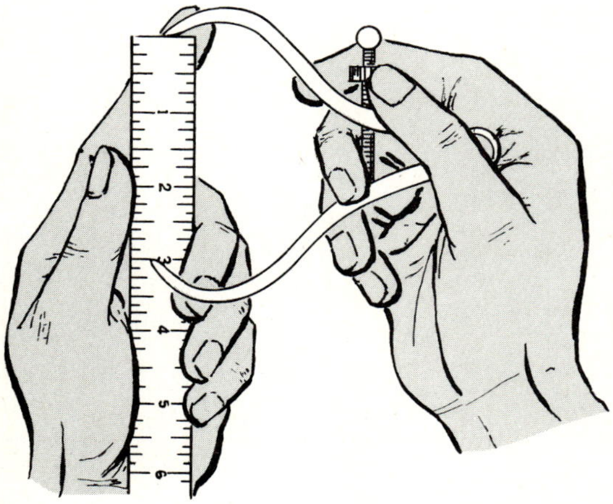

Fig. 2-4. Checking the measurement to which the outside caliper has been adjusted. (*South Bend Lathe Works*)

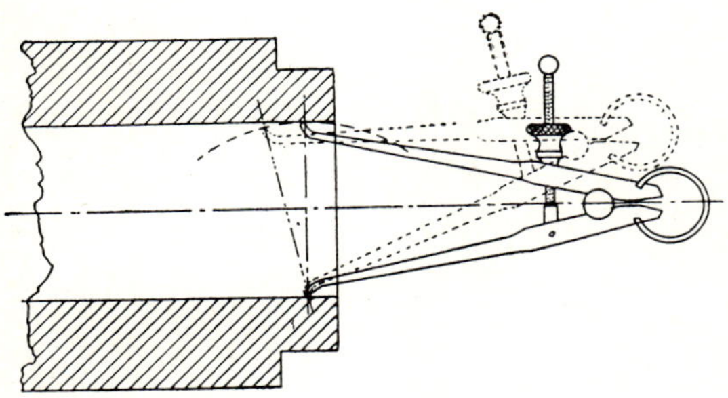

Fig. 2-5. Adjusting inside calipers to the size of a hole.

MEASUREMENTS WITH INSIDE CALIPER	
Piece 1	______ in
Piece 2	______ in
Piece 3	______ in
Piece 4	______ in
Piece 5	______ in

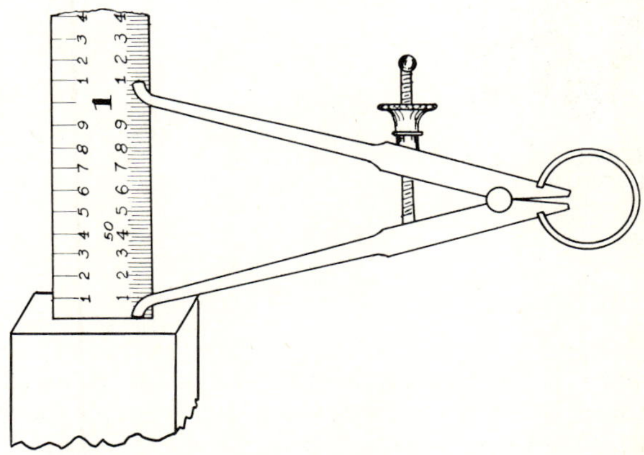

Fig. 2-6. Measuring the distance between the legs of the caliper.

_____ **6** Steel rules and calipers should be kept clean and dry. An occasional wiping, with a clean cloth slightly dampened with oil, will prevent rusting. Do not damage them by throwing them in your toolbox or by piling other tools on top of them.

True-or-False Test — Job 2

1 A steel scale can be used to measure the diameter of a hole. **1** ______

2 The smallest mark on a rule or steel scale is usually $\frac{1}{4}$ in. **2** ______

3 You could not use a steel scale to make a measurement as small as 0.001 in. **3** ______

4 You can measure the outside diameter of a shaft using an outside caliper and a steel scale. **4** ______

5 You can measure the inside diameter of a hole using an inside caliper and a steel scale. **5** ______

6 When using an outside caliper, you tighten the thumb nut until the legs can just barely be forced on the shaft. 6 ______

7 You can clean steel scales and calipers by wiping them with a clean cloth slightly dampened with oil. 7 ______

8 Some steel rules are marked off in decimal fractions. 8 ______

9 $\frac{1}{4}$ in is equal to 0.41 in. 9 ______

10 0.375 in is equal to $\frac{3}{8}$ in. 10 ______

Written Checkup Job 2

1 Write down the procedure for measuring the diameter of a shaft with calipers and rule.

2 Write down the procedure for measuring the diameter of a hole with calipers and rule.

3 Use Table 2-1 to fill in the spaces below. Translate decimals to fractions or fractions to decimals, as required.

$\frac{1}{4}$ in is ______ (decimal)

$\frac{5}{8}$ in is ______ (decimal)

$\frac{13}{16}$ in is ______ (decimal)

$3\frac{7}{8}$ in is ______ (decimal)

0.2344 in is ______ (decimal)

0.875 in is ______ (fraction)

2.2969 in is ______ (fraction)

0.9843 in is ______ (fraction)

NAME DATE CLASS

Objective: To learn about the metric system of measurement, and how to use it.

Equipment: Round stock or shafts, retangular stock, plates or machine parts with holes in them.

Tools: Metric rule.

Job 3

THE METRIC SYSTEM

PROCEDURE

Part One SYSTEMS OF MEASUREMENTS

_____ **1** The two most widely used measuring systems are the U.S. Customary System and the metric system. The U.S. Customary System (USCS) uses inches, feet, miles, pints, quarts, gallons, and so on. This is the system of measurement we worked with in Job 2. The United States is the only major country in the world still using the USCS. Most countries already use the metric system or are in the process of switching to it.

_____ **2** Cars, motorcycles, and machinery imported into the United States are dimensioned in the metric system. Also, some engines built in the United States are already completely metric. This means that all measurements, all nuts and bolts, are in metric units. If you work on cars and other machinery from foreign countries, you will need an extra set of tools—metric tools. In the meantime, the United States is gradually switching to the metric system. This system has already been adopted by some industries, trade organizations, and professional societies. Let's look at the U.S. Customary System, and then discuss the metric system.

_____ **3** Our familiar USCS measurements originated hundreds of years ago in England. They were brought to this country by the first settlers.

_____ **4** While many colonies in America followed the English system of measurement, some did not. Different standards of measure among the colonies made trade difficult. The Articles of Confederation in 1781 and the U.S. Constitution in 1790 gave Congress the power to fix uniform standards for weights and measures. The effect of this action was to create what we know as the U.S. Customary System. It is almost the same today as it was when first used by the early colonists.

_____ **5** Measurements in the U.S. Customary System include:

LENGTH

12 inches = 1 foot
3 feet = 1 yard
5,280 feet = 1 mile

LIQUID

16 ounces = 1 pint
2 pints = 1 quart
4 quarts = 1 gallon

WEIGHT

16 ounces = 1 pound
2,000 pounds = 1 ton

_____ **6** Look at all these figures, and at the different units of measure. They could be confusing, especially to someone who is not familiar with them. Yet these are only some of the figures and units a person must learn when working with the U.S. Customary System.

_____ **7** Not all the world was satisfied with the English system of measure. Many people thought it was too complicated, because the units have little relation to each other. Many other people did not use the English system at all. They preferred units that were peculiar to their own country or locality.

_____ **8** In 1670, a Frenchman named Gabriel Mouton recognized the advantages of a worldwide measurement system. He proposed a decimal measurement system. It would be based on the length of 1 minute of arc of a great circle of the earth. Few people listened to Mouton.

_____ **9** However, during the next 100 years, people gradually became aware of the need for a standard measurement system for all countries. During the French Revolution in 1790, the National Assembly of France asked the French Academy of Sciences to "deduce an invariable standard for all the measures and all the weights." A commission appointed by the Academy created what we know today as the metric system of measurement.

Part Two UNITS OF THE METRIC SYSTEM

_____ **1** In use, the metric system proved to be both scientific and simple. The simplicity of the system is its use of multiples of 10. The larger and smaller units are obtained by multiplying or dividing the basic units by 10 and its multiples. Multiplication or division by 10 is performed simply by moving the decimal point.

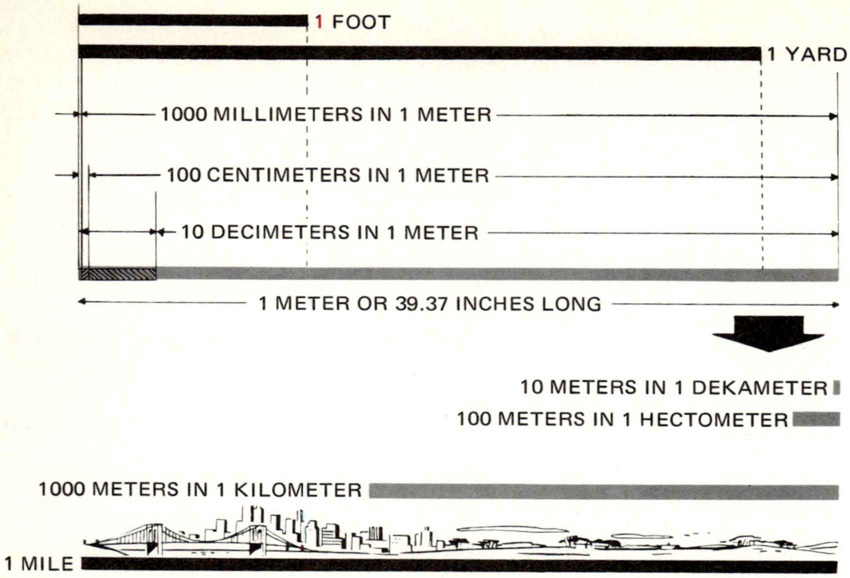

Fig. 3-1. The meter is the metric unit of length. (*General Motors Corporation*)

_____ **2** The unit of length is called the *meter* (Fig. 3-1). The name came from the Greek word *metron,* meaning "a measure." As you can see in Fig. 3-1, a meter is equal to 39.37 in. In other words, a meter is slightly longer than a yard.

_____ **3** The metric unit of mass or weight is called the *gram* (Fig. 3-2). It is the mass of 1 cubic centimeter of water at its temperature of maximum density.

_____ **4** A *cubic centimeter* is a cube that is $\frac{1}{100}$ of a meter on each side. The *cubic decimeter* (a cube $\frac{1}{10}$ of a meter on each side) was chosen as the unit of fluid capacity. This measure was given the name *liter* (Fig. 3-2). The liter is slightly smaller than the quart. Both cubic centimeters and liters are used to indicate the piston displacement or size of an engine. (In the USCS, cubic inches are used.)

_____ **5** As you can see from our brief discussions, the USCS just grew for several hundred years. It did not grow logically and, therefore, did not lend itself to scientific and industrial use.

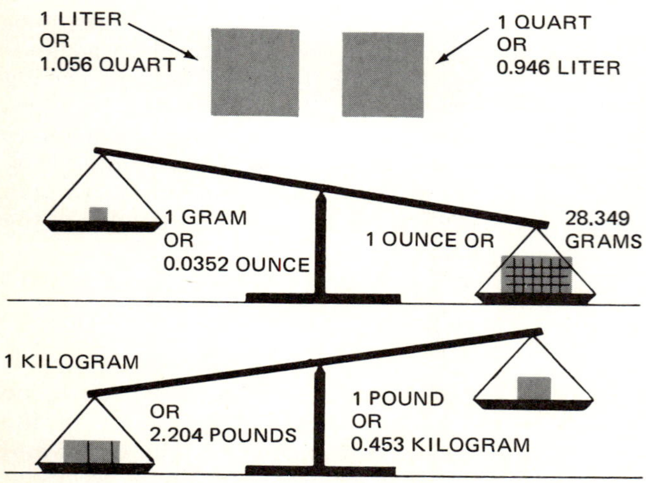

Fig. 3-2. The gram is the metric unit of mass, and the liter is the metric unit of fluid capacity or volume. (*General Motors Corporation*)

_____ **6** In contrast, the metric system is scientific (Fig. 3-3). It is based on units that are divided into multiples of 10, just as our monetary system is based on dividing the dollar into 100 parts. In the metric system, for example, 1 meter (which is 39.37 in) is 10 decimeters, or 100 centimeters, or 1,000 millimeters.

_____ **7** As you can see in Fig. 3-3, three metric-system units replace many different USCS units. Weight is measured in grams, length in meters, and volume in liters.

_____ **8** In changing to the metric system, the United States would not have to change its money system. That is already based on the decimal system. Also, time is measured in the same units—hours, minutes, and seconds—used in the U.S. Customary System. Now, let's look more closely at some basic metric measurements.

_____ **9** Measurements in the metric system include:

LENGTH

1 millimeter = 0.039 in
10 millimeters = 1 centimeter (0.394 in)
100 centimeters = 1 meter (39.37 in, or a little more than a yard)
1,000 millimeters = 1 meter
1,000 meters = 1 kilometer (0.62 mile)

VOLUME

1,000 milliliters = 1 liter (1.057 quarts)

WEIGHT

1,000 milligrams = 1 gram (0.035 ounce, or the approximate weight of a paper clip)
1,000 grams = 1 kilogram (2.2 pounds)
1,000,000 milligrams = 1 kilogram
1,000 kilograms = 1 metric ton

_____ **10** You have probably noticed that the names of metric units sometimes include prefixes (milli, centi, kilo, etc.) as in milliliter, centimeter, and kilogram. These prefixes indicate multiples or submultiples of the units. Here are the most commonly used prefixes, and the multiplication factors they indicate.

PREFIX	MULTIPLICATION FACTOR
kilo	1,000 (one thousand)
centi	0.01 (one hundredth)
milli	0.001 (one thousandth)

_____ **11** The term kilometer means 1,000 meters. A centimeter is $\frac{1}{100}$ of a meter. A millimeter is $\frac{1}{1000}$ of a meter. The conversion table (Table 3-1) gives you the measurement equivalents you will need to know to work on imported cars and motorcycles. With this table, you can convert USCS measurements to metric measurements, or metric measurements to USCS measurements. Table 3-1 also gives the standard abbreviations for USCS and metric measurements. We shall use these abbreviations in the rest of this workbook.

_____ **12** In most service manuals for imported cars and motorcycles, the manufacturers give specifications in both USCS and metric measurements. For example, the piston stroke of an engine might be

TABLE 3-1

	LENGTH	
1 in (inch)	= 2.54 cm (centimeters)	or 0.0254 m (meter)
1 cm	= 0.39 in	or 0.03281 ft (foot)
1 mm (millimeter)	= 0.039 in	or 0.003281 ft
1 ft	= 30.48 cm	or 0.3048 m
1 mi (mile)	= 1.609 km (kilometers)	or 1,609 m
1 km	= 0.62 mi	or 3,281 ft
	VOLUME/CAPACITY	
1 in^3 (cubic inch)	= 16.39 cm^3 (cubic centimeters)	or 0.01 liter
1 cm^3	= 0.061 in^3	or 0.001 liter
1 liter	= 61.02 in^3	or 1.057 qt (quarts)
1 gal (gallon)	= 4 qt	or 3.780 liters
	WEIGHT	
1 kg (kilogram)	= 2.2 lb (pounds)	= 35.2 oz (ounces)
1 lb	= 0.454 kg	

Here are the metric measurements, taken from the complete metric-system table, that you will work with most often.

	LENGTH	
1 km	= 1,000 m	= 100,000 cm
1 m	= 100 cm	= 1,000 mm (millimeters)
	VOLUME/CAPACITY	
1 kl (kiloliter)	= 1,000 liters	= 100,000 cl (centiliters)
1 liter	= 1,000 cm^3	= 1,000 ml (milliliters)
	WEIGHT	
1 kg (kilogram)	= 1,000 g (grams)	= 100,000 cg (centigrams)

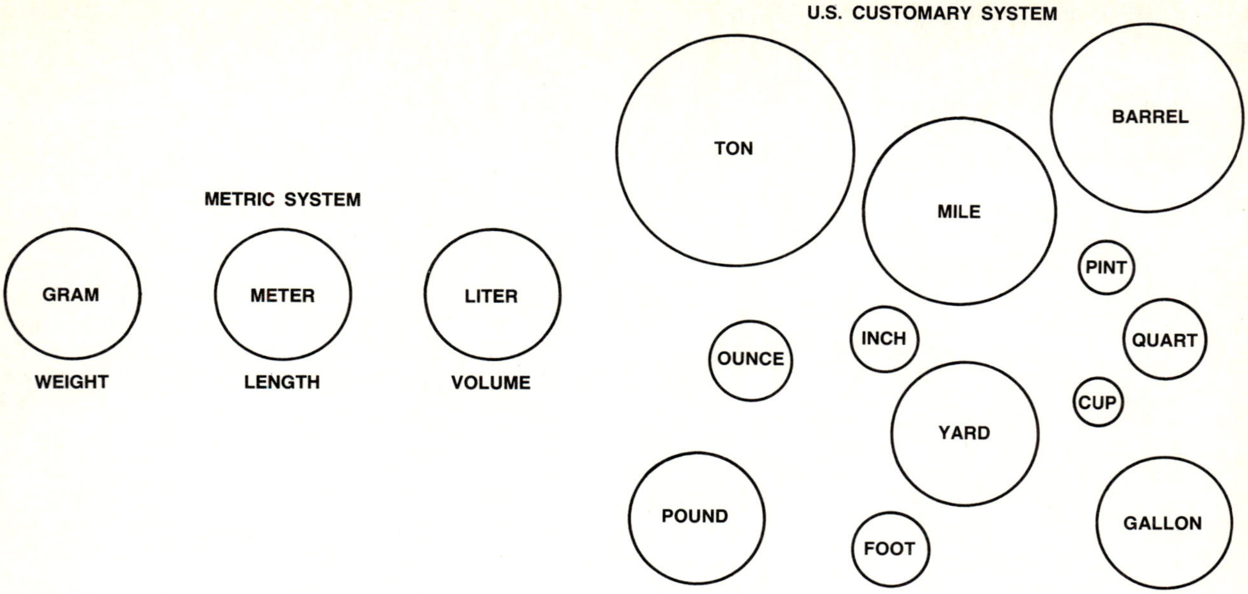

Fig. 3-3. Three metric units replace many different U.S. Customary System units.

given as 73.7 mm (millimeters) [2.90 in]. The torque for tightening main-bearing cap bolts might be given as 4.5 to 5.5 kg-m (kilogram-meters) [32.5 to 39.5 lb-ft (pound-feet)].

____ **13** To work on imported cars and motorcycles, you will need a set of metric wrenches and sockets. We shall discuss metric hand tools in Part 3 of this workbook. USCS-type tools will not fit some nuts and bolts in imports.

____ **14** Many mechanics work with the metric system regularly. They memorize the frequently used measurements. Others find the metric equivalent by working a separate arithmetic problem for each specification. Such a process is slow, and mistakes can be made easily.

____ **15** Metric equivalents can be found much faster and more accurately with conversion tables like those in the appendix at the end of this workbook. In Part 6 we shall work some exercises, to familiarize you with the tables. These tables can save you time in determining metric equivalents.

Part Three METRIC RULES

____ **1** Figure 3-4 shows a rule or steel scale that is marked in both inches (United States Customary System) and centimeters (metric system). This is called *dual dimensioning.* If you have to buy a steel scale, try to find one that is dual-dimensioned. That is, buy a rule with inches on one edge and centimeters on the other.

____ **2** Now, obtain the five pieces of stock, or objects, that you measured with the steel scale in Job 2. If you have a steel rule with a metric scale, measure each piece again. Write down its metric measurement in the blanks in the following table. Enter each measurement in millimeters (mm).

MEASUREMENTS WITH A STEEL RULE

Piece 1	______ mm
Piece 2	______ mm
Piece 3	______ mm
Piece 4	______ mm
Piece 5	______ mm

____ **3** Measurements made in inches in the U.S. Customary System can be converted to metric measurements by using Table 3-1. In the left column below, write the inch measurements you obtained with a steel rule in Part Two of Job 2. Then, using Table 3-1, convert each inch measurement to its metric equivalent. Write the results in the blanks on the right.

METRIC-CONVERSION PRACTICE

Piece 1	______ in	______ mm
Piece 2	______ in	______ mm
Piece 3	______ in	______ mm
Piece 4	______ in	______ mm
Piece 5	______ in	______ mm

____ **4** Now check up on yourself to see how well you understand and can work with the metric system. In Step 2 you made metric measurements and wrote them in the blanks. In Step 3 you used arithmetic and the conversion table to convert from inches to millimeters. So the measurement you have for Piece 1 in Step 2 should be the same as the metric

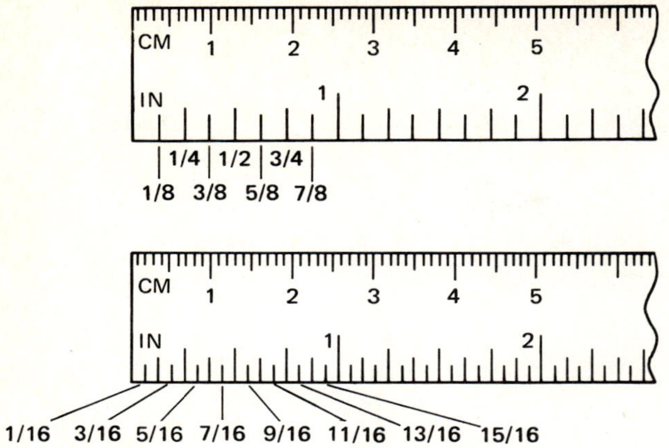

Fig. 3-4. Rule, or steel scale, marked in both inches (United States Customary System) and centimeters (metric system).

result you have for Piece 1 in Step 3. This should also be true for the other measurements and results. Compare them to see how close you came.

Multiple-Choice Test — Job 3

1 In the metric system, the unit of length is the **A** foot **B** meter **C** liter **D** gram **1** ______

2 The metric unit of measure that is slightly longer than a yard is the **A** meter **B** millimeter **C** kilometer **D** liter **2** ______

3 In the metric system, weight is measured in **A** liters **B** meters **C** cubic centimeters **D** grams **3** ______

4 The metric measure of volume that is slightly larger than a quart is the **A** liter **B** meter **C** gram **D** volt **4** ______

5 A steel scale, rule, or other measuring instrument that has both metric and USCS scales is called **A** a ruler **B** a caliper **C** dual-dimensioned **D** a unit of measure **5** ______

Written Checkup — Job 3

1 Explain how the United States Customary System and the metric system differ.

2 Study the conversion table until you understand how to convert inches to millimeters, millimeters to inches, miles to kilometers, gallons to liters, and so on. Then make the following conversions:

1 in = ______ mm
1 mm = ______ in
1 gal = ______ liters
1 liter = ______ gal
1 mi = ______ km
1 km = ______ mi
1 kg = ______ lb
1 lb = ______ kg
1 in = ______ cm
0.039 in = ______ mm

NAME DATE CLASS

Objective: To learn how to use feeler gauges.

Equipment: Five pieces or parts with gaps to be measured, five used spark plugs.

Tools: Various sets and types of feeler gauges.

Job 4

FEELER GAUGES

PROCEDURE

_____ **1** Feeler gauges are strips or blades of hardened steel or other metal. They are ground or rolled to the proper thickness with extreme accuracy. This is the reason feeler gauges are sometimes called "thickness gauges." They are generally supplied in sets (Fig. 4-1). Each blade in the set is marked with its thickness in thousandths of an inch, thousandths of a millimeter, or both. For example, in Fig. 4-1, the gauge marked 0.003 in is also marked with its metric equivalent of 0.08 mm. Figure 4-2 shows a feeler gauge being used to check piston ring clearance.

_____ **2** Some feeler gauges have two steps, or thicknesses. These are *stepped* feeler gauges (Fig. 4-3). The tip of the blade of a stepped feeler gauge is thinner than the rest of the blade (Fig. 4-4). The blade

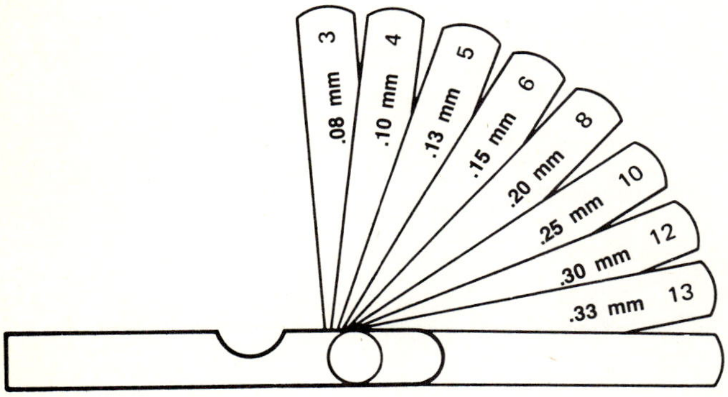

Fig. 4-1. Set of feeler gauges.

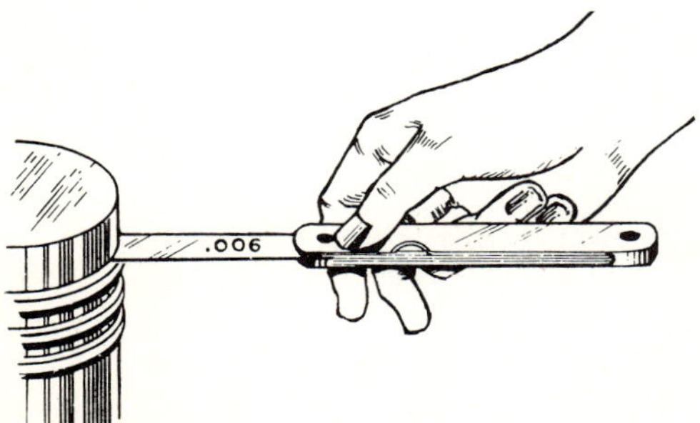

Fig. 4-2. Checking piston-ring clearance with a feeler gauge.

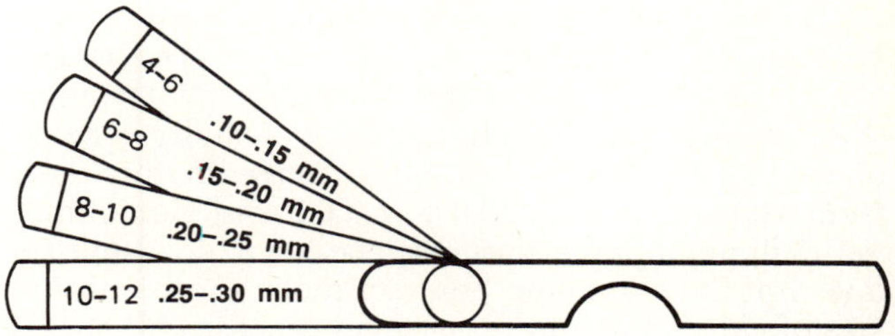

Fig. 4-3. Set of stepped feeler gauges.

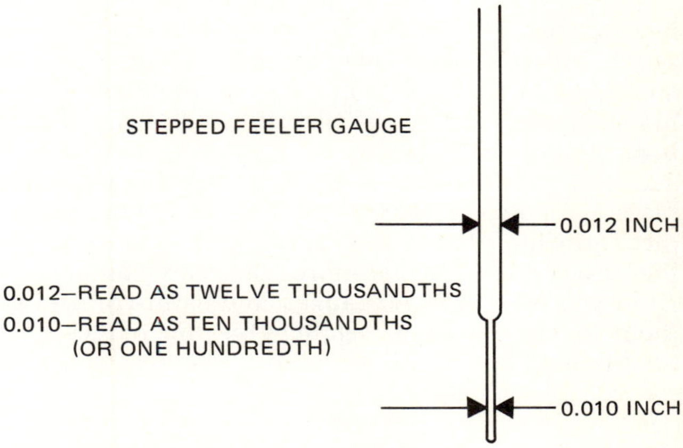

Fig. 4-4. Blade of a stepped feeler gauge. The tip is thinner than the rest of the blade.

marked 4–6 in Fig. 4-3 is 0.004 in [0.10 mm] thick at the tip, and 0.006 in [0.15 mm] thick on the thicker portion that starts about $\frac{1}{2}$ in back from the end of the blade.

_____ **3** Stepped feeler gauges are handy on certain jobs. For example, the specifications might call for a clearance of 0.005 in [0.13 mm]. With the 0.004–0.006 in [0.10–0.15 mm] gauge, the clearance is adjusted so that the 0.004-in [0.10-mm] portion fits but the 0.006-in [0.15-mm] portion does not fit. For this reason, stepped feeler gauges are often called "go no-go" gauges.

_____ **4** Wire feeler gauges (Fig. 4-5) are similar to flat feeler gauges, but they are made of carefully calibrated steel wire of the proper thickness. They are useful in checking spark-plug gaps and similar dimensions (Fig. 4-6). Metric wire feeler gauges are also available. When specifications are given in the

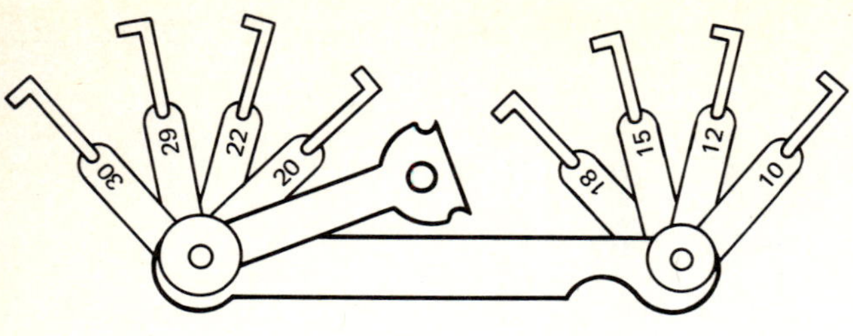

Fig. 4-5. Set of wire feeler gauges.

metric system, and you do not have metric gauges, you can convert the measurements with Table 3-1.

_____ **5** Feeler gauges are made in many shapes and sizes. Some feeler gauges, or blades, are straight, as in Figs. 4-1 to 4-3. Others are bent at an angle (Figs. 4-5 and 4-6). A feeler-gauge set usually consists of several blades, which often range in thickness from 0.0015 to 0.040 in [0.037 to 1.02 mm]. Sometimes you will need to measure a space for which your set does not have a blade. You can make up the proper size of gauge by combining several blades. The total thickness of all the blades used to fill the space is the measurement between the surfaces.

_____ **6** Feeler gauges are precision measuring instruments. They must be treated with care to preserve their accuracy. Never force a feeler gauge into an opening that is too small for the thickness of the blade. Some blades are very thin; they are easily bent, dented, and torn.

_____ **7** Before using a feeler gauge, wipe the blade with a clean oiled cloth. This will remove any dirt from the blades, and prevent inaccurate readings (especially if two or more blades are combined). After you use the feeler gauges, and before returning them to your toolbox, again wipe the blades and holder with a clean oiled cloth. This prevents the acids and moisture on your hands from rusting the blades.

_____ **8** Most feeler gauges are made of steel. However, sometimes a mechanic needs a nonmagnetic feeler gauge (Fig. 4-7). Brass feeler gauges are available for work around permanent magnets. They are used, for example, to measure the air gap in an electronic distributor. The permanent magnet in the distributor attracts the steel gauge and prevents accurate measurement.

_____ **9** Many engine tuneup kits contain a plastic disposable or throwaway-type feeler gauge. Figure 4-8 shows such a gauge and how it is used. Note that, on this particular gauge, one end is a round wire gauge for use in gapping the spark plugs. The other end is a flat feeler gauge for use in gapping the contact points in the ignition distributor.

_____ **10** Examine various feeler-gauge sets. Notice the difference between the plain, flat-blade type of feeler gauge and the stepped feeler gauge (Figs. 4-1 and 4-3).

_____ **11** Next, compare the flat-blade type with a set of round wire feeler gauges (Figs. 4-3 and 4-5). Note how the stepped feeler gauge is marked to indicate the thickness of each part of the blade.

_____ **12** Now, obtain five pieces or parts which have clearances or gaps to be measured. Using a standard feeler-gauge set, take each measurement. Write it down in the left column below. Then, using Table 3-1, figure the metric measurement for each gap. Write the metric equivalent in the proper place in the right column below.

MEASUREMENTS WITH FEELER GAUGES

Piece 1	_____ in	_____ mm
Piece 2	_____ in	_____ mm
Piece 3	_____ in	_____ mm
Piece 4	_____ in	_____ mm
Piece 5	_____ in	_____ mm

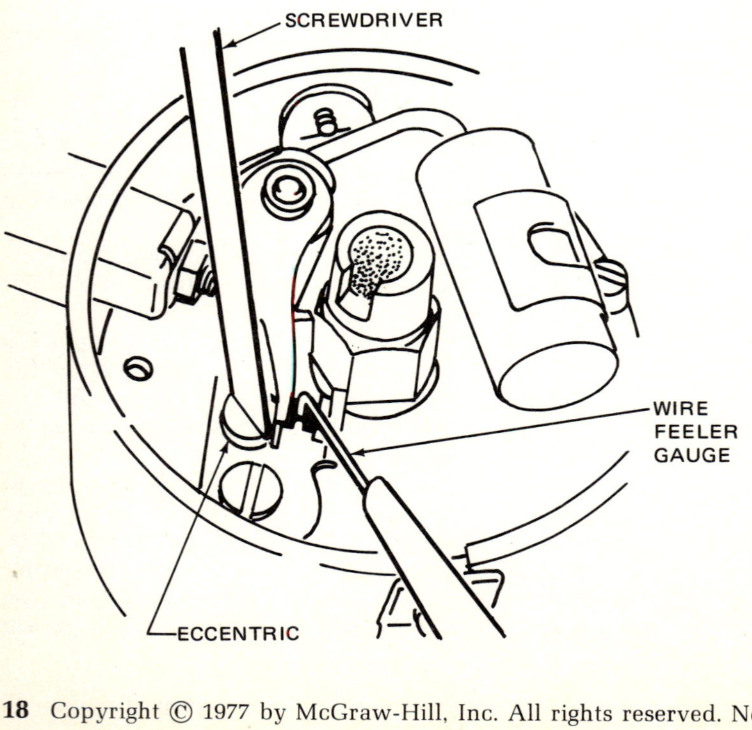

Fig. 4-6. Checking distributor contact-point opening, or gap, with a wire feeler gauge. (*Delco-Remy Division of General Motors Corporation*)

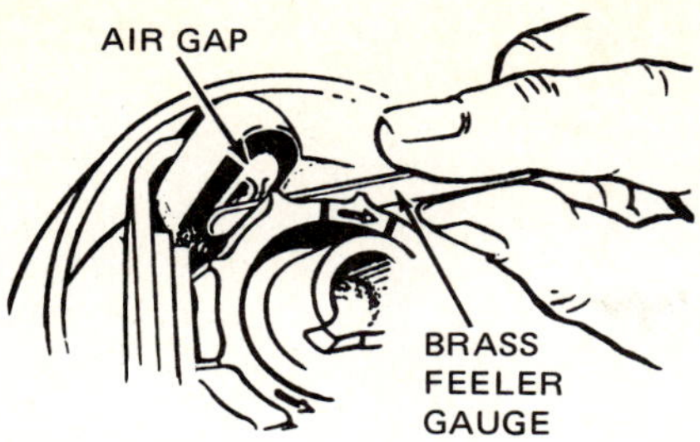

Fig. 4-7. Checking the reluctor air gap in an electronic distributor using a nonmagnetic feeler gauge. (*Echlin Manufacturing Company*)

_____ **13** Using a set of stepped feeler gauges, measure each gap or clearance again. Write the measurement in the left column below. (Remember how to use a stepped feeler gauge: If the tip of the blade goes into the gap, and the bigger part of the blade does not, the measurement is halfway between the two thicknesses of the blade.) Then convert each measurement to its metric equivelent. Enter that in the right column.

MEASUREMENTS WITH STEPPED FEELER GAUGES

Piece 1	_____ in	_____ mm
Piece 2	_____ in	_____ mm
Piece 3	_____ in	_____ mm
Piece 4	_____ in	_____ mm
Piece 5	_____ in	_____ mm

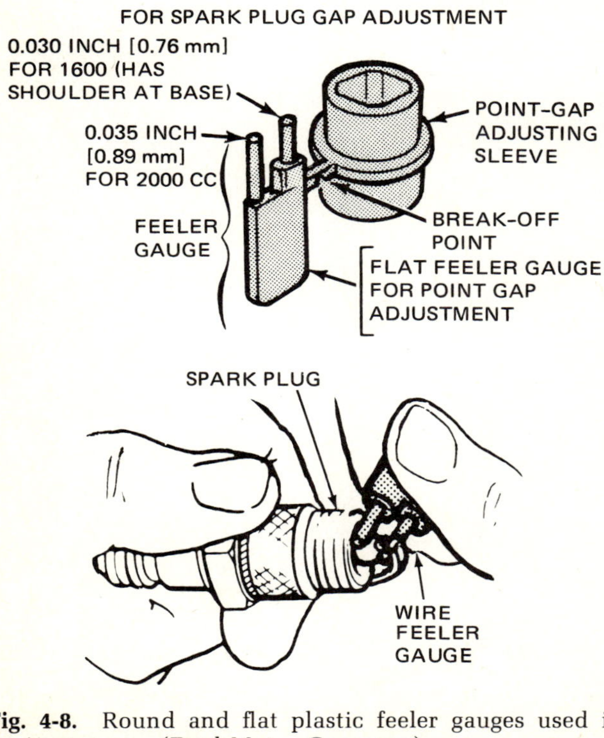

Fig. 4-8. Round and flat plastic feeler gauges used in engine tuneup. (*Ford Motor Company*)

_____ **14** Now, use the wire feeler gauges to measure the gaps of five used spark plugs. Write the measurements in the left column. Next measure the gaps of the same five used spark plugs with the flat feeler gauges. Write these measurements in the right column.

MEASURING SPARK-PLUG GAPS

	WIRE FEELER GAUGE	FLAT FEELER GAUGE
Piece 1	_____ in	_____ in
Piece 2	_____ in	_____ in
Piece 3	_____ in	_____ in
Piece 4	_____ in	_____ in
Piece 5	_____ in	_____ in

True-or-False Test — Job 4

1 Feeler gauges can be used to measure piston ring clearance. **1** _____

2 Feeler gauges are always the same thickness from one end to the other. **2** _____

3 Wire, or round, feeler gauges are made of copper wire. **3** _____

4 A stepped feeler gauge has two thicknesses. **4** _____

5 If a feeler gauge becomes rough or bent, it can be used again after it is hammered out smooth. **5** _____

6 The air gap in an electronic distributor is measured with a nonmagnetic feeler gauge. **6** _____

7 Some round and flat feeler gauges are made of plastic. **7** _____

8 Metal feeler gauges are maintenance-free, and do not require cleaning or oiling. **8** _____

9 You can buy feeler gauges marked in thousandths of an inch and in thousandths of a gram. **9** _____

10 Wire feeler gauges are preferred for checking electrical gaps, such as spark-plug gaps. **10** _____

Written Checkup — Job 4

1 Describe the general procedure for using feeler gauges.

2 In Step 14 you measured five spark-plug gaps with wire feeler gauges and with flat feeler gauges. These measurements may not be the same. The flat-feeler-gauge measurements are probably smaller than the wire-feeler-gauge measurements. Describe the reason for the differences in the measurements.

NAME DATE CLASS

Objective: To learn how to use micrometers.
Equipment: Round and flat stock, engine block.
Tools: Outside and inside micrometers.

Job 5
MICROMETERS

Safety: *Never attempt to measure a part that is moving—a shaft that is turning, for example. The micrometer might jam on the shaft, be whirled around, and break. If this happens, you might be seriously injured by flying particles.*

PROCEDURE

Part One READING A MICROMETER

_____ **1** Measurements made with a caliper and scale (Job 2) are usually not accurate enough for engine service and repair work. In such work, measurements must be much finer than $\frac{1}{64}$ in [0.397 mm]. For these measurements, more precise instruments are used, such as the micrometer, dial indicator, and vernier caliper.

_____ **2** One special form of caliper is called a *micrometer* (Fig. 5-1). The micrometer measures much more accurately than an ordinary caliper. In addition, the measurements can be read directly from the micrometer, without using a separate steel scale. Instead of reading in fractions of an inch like many rules, the micrometer reads in tenths, hundredths, thousandths, and sometimes ten-thousandths of an inch. That is, it uses the decimal system.

_____ **3** In the shop, a micrometer is usually called a "mike," and measuring something with the mike is called "miking." To use the micrometer to mike the diameter of a round shaft or stock, you hold it as shown in Fig. 5-2.

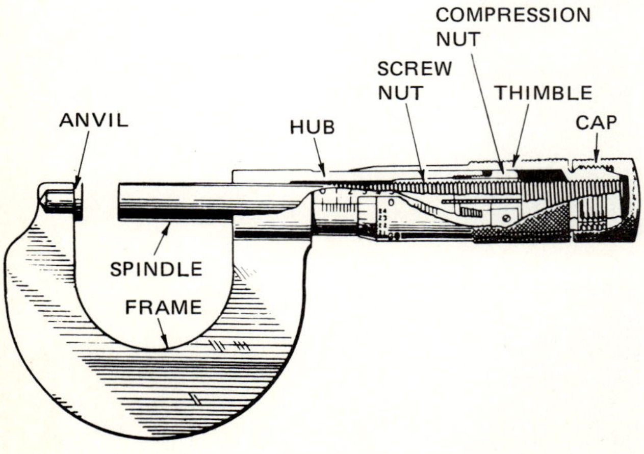

Fig. 5-1. Outside micrometer.

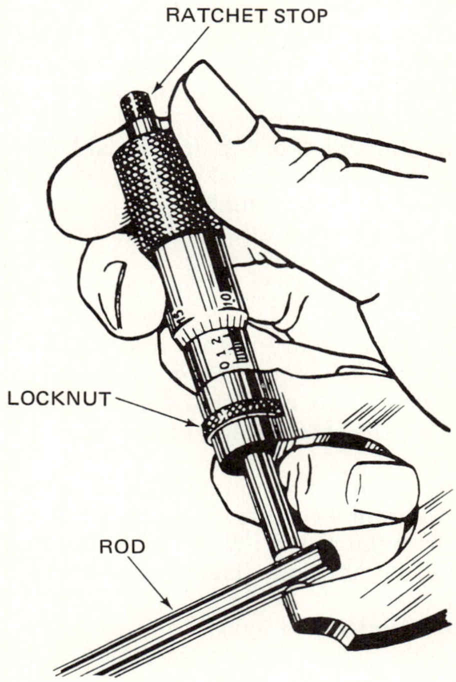

Fig. 5-2. Using an outside micrometer to measure the diameter of a rod.

_____ **4** Examine an outside micrometer. Identify the anvil, frame, spindle, hub, and thimble (Fig. 5-1).

In Fig. 5-2, note the locknut and the ratchet stop. Here's how the locknut is used. Suppose, for example, that you measured a shaft, but that the position of the micrometer did not allow you to see the reading. Then, you could turn the locknut with your fingers. This would lock the spindle at the final setting, so you could remove the micrometer from the shaft and take the reading.

The ratchet stop (Fig. 5-2) is a friction clutch that is built into the micrometer. By turning the ratchet stop instead of the thimble, when the proper pressure is applied by the anvil to the piece being measured, the ratchet stop will slip. This prevents the spindle from turning any further. Use of the ratchet stop makes the taking of readings faster and more accurate, especially for someone not accustomed to using the micrometer.

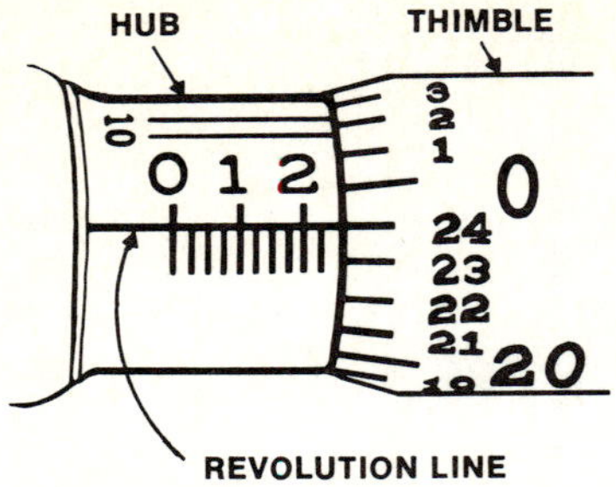

General Motors Corporation

Fig. 5-3. Tub and thimble markings on a micrometer.

_____ **5** Turn the thimble so that the spindle moves away from the anvil. Note the markings on the hub and the thimble. The measurement is read off the hub and thimble.

_____ **6** Note, in Fig. 5-3, that the hub (or barrel) is marked off in uniform spacings of 0.025 in (twenty-five thousandths of an inch).

_____ **7** Note that the circumference of the thimble is marked off by 25 graduations. Each graduation represents 0.001 in (one thousandth of an inch).

_____ **8** To measure with a micrometer, turn the thimble until the piece being measured is a light-drag fit between the anvil and the end of the spindle. Then read the measurement on the barrel and thimble.

_____ **9** Each barrel marking indicates 0.025 in. Four of them make 0.1 in. If three of the figures are visible, as at the top in Fig. 5-4, then the thimble has uncovered 12 markings. The measurement is at least 0.300 in.

_____ **10** Notice the words "at least" in Step 9. Checking the barrel is only the first step in reading the measurement. The second step is to note exactly where the markings on the thimble are, relative to the line on the barrel.

_____ **11** At the top in Fig. 5-4, for instance, the thimble has been turned so that its marking number 4 lines up with the barrel line. This means that the thimble has been turned 0.004 in past the 0.300-in mark. Each thimble marking represents 0.001 in. The actual measurement being shown at the top is 0.304 in (0.300 plus 0.004).

_____ **12** Now, look at the middle and the bottom of Fig. 5-4. In the middle, the thimble has been turned just one thimble marking past the ninth barrel marking. The barrel markings indicate 0.225 in. The single thimble marking, indicating 0.001 in, must be added to give an actual measurement of 0.226 in.

_____ **13** As you can see, you get a mike reading as follows: First note how many barrel markings were exposed as the thimble was backed off. Each barrel marking represents 0.025 in. Then, notice which thimble marking is aligned with the barrel line. Each thimble marking represents 0.001 in. Add the two to get the measurement.

_____ **14** Now, notice the bottom measurement in Fig. 5-4. Here the thimble has been turned to uncover only eight barrel markings, so the reading is 0.200 in. But the thimble has been turned until its 24 marking lines up with the barrel line. The thimble thus shows 0.024 in. The total measurement is therefore 0.224 in (0.200 plus 0.024.)

Part Two THE METRIC MICROMETER

_____ **1** The metric micrometer (Fig. 5-5) reads in millimeters (mm) and hundredths of a millimeter (0.01 mm). The barrel is marked off in millimeters

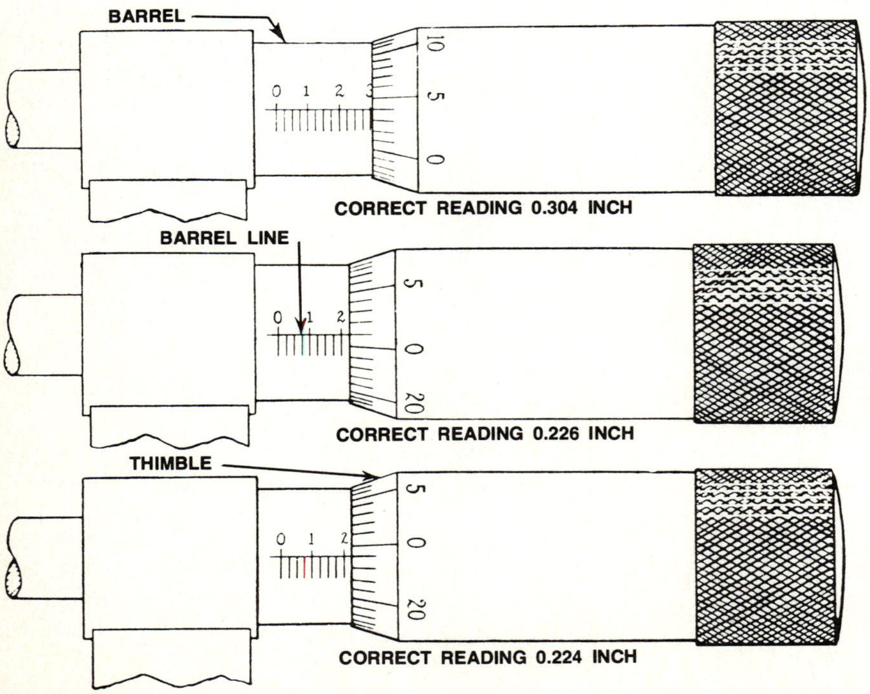

Fig. 5-4. Reading a micrometer.

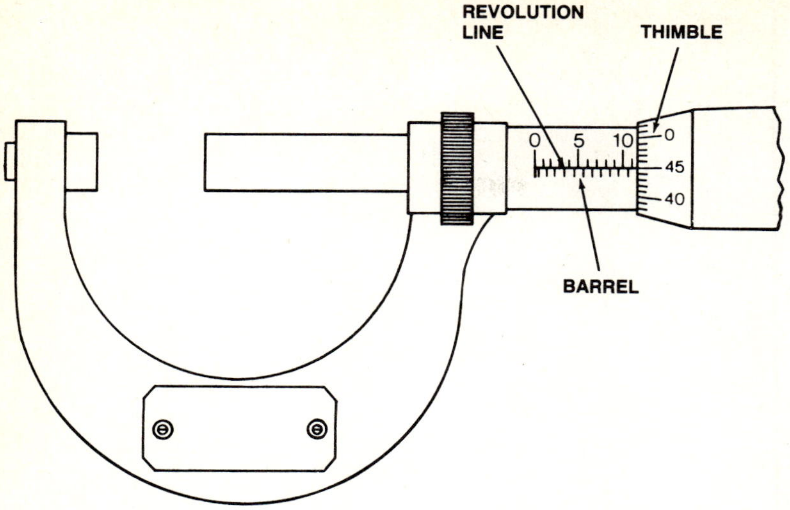

Fig. 5-5. Metric micrometer. (*Volkswagen of America, Inc.*)

above the line, and half millimeters (0.50 mm) below the line. The thimble is marked off into divisions of 0.01 mm.

_____ **2** To read the micrometer in Fig. 5-5, add the reading on the barrel (11 mm) to the reading on the thimble (0.45 mm). The total measurement is 11.45 mm. As you can see, reading a metric micrometer is probably easier than reading a mike marked in inches.

Part Three USING A MICROMETER

_____ **1** The micrometer has very precisely cut internal screw threads. Rough treatment will ruin it. A micrometer should not be thrown about carelessly on the workbench. It should be kept in a special drawer or container, where it is protected from dirt and from other tools being dropped on it. It should be wiped clean after every use.

_____ **2** Never clamp the micrometer on the piece being measured. The thimble should be tightened only enough to cause the micrometer to drag slightly as it slides over the piece. Clamping distorts the threads and frame, and could ruin the micrometer.

_____ **3** Never try to measure a piece that is turning in a lathe. The micrometer might tighten on the piece and be torn out of your hand. This could ruin the work and the micrometer, and injure you seriously.

_____ **4** Use the outside micrometer to measure the thickness or diameter of a piece of stock (Fig. 5-2) or a blade from a feeler-gauge set. Turn the thimble back enough so that the anvil and spindle surfaces pass over the stock without interference. Then turn the thimble slowly forward (clockwise), while repeatedly passing the micrometer over the stock. Continue until the micrometer drags slightly as it passes over the piece being measured.

_____ **5** After adjusting the micrometer to the correct fit, read the measurement. Note first the full-inch mark or micrometer size; that is, the number of complete inches (0, 1, 2, etc.). Then, count the exposed markings on the hub, remembering that each is 0.025 in. Finally, note the marking on the thimble that is aligned with the heavy line on the hub. Adding these together gives the dimension of the piece that was miked. Write the measurement in the first space below. Then measure other objects, and record the readings.

MEASUREMENTS WITH AN OUTSIDE MICROMETER	
Piece 1	_______ in
Piece 2	_______ in
Piece 3	_______ in
Piece 4	_______ in
Piece 5	_______ in

_____ **6** Many variations of the micrometer are used in shop work. For example, there are inside micrometers (Fig. 5-6) for measuring hole diameters (Fig. 5-7).

_____ **7** In addition, many of the precision power machines in the shop have micrometer adjustments. Cylinder honing or boring equipment, lathes, and grinders have micrometer adjustments so that the machining operations can be carefully controlled.

_____ **8** A recent development is the digital micrometer (Fig. 5-8). This micrometer makes mike reading faster and easier. The actual measurement appears automatically in the "counter" on the frame.

_____ **9** Figure 5-8 shows a digital micrometer that reads in the metric system. The measurement that appears in the counter is in millimeters.

Part Four USING AN INSIDE MICROMETER

_____ **1** Inside micrometers (Fig. 5-6) are used to measure hole diameters, such as the diameter, or bore, of an engine cylinder. (See Fig. 5-7.) Extension rods, such as those above the inside micrometer in Fig. 5-6, can be attached so the mike will measure large diameters.

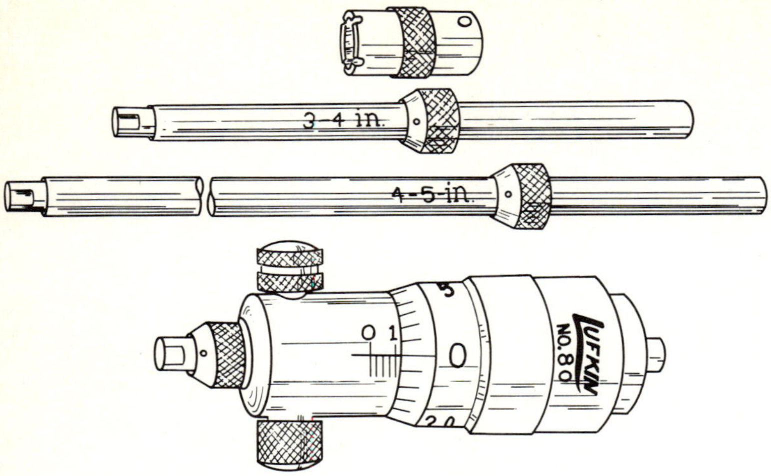

Fig. 5-6. Inside micrometer.

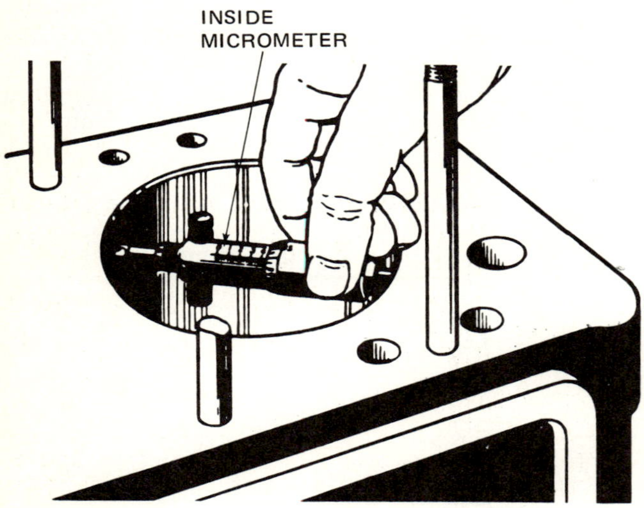

Fig. 5-7. Using an inside micrometer to measure the diameter, or bore, of a cylinder.

_____ **2** Examine an inside micrometer. Note the markings on the hub and thimble. They correspond to the markings on the hub and thimble of the outside micrometer (Fig. 5-3). The extension rods allow you to measure cylinder diameters from $1\frac{1}{2}$ in to several inches.

_____ **3** When inserting an extension rod in an inside micrometer, be sure that the extension seats properly. The shoulder on the rod must seat fully for the micrometer to read correctly.

_____ **4** Figure 5-7 shows an inside micrometer being used to measure the bore of an engine cylinder. To measure a cylinder bore, first turn the thimble clockwise until the length of the assembled micrometer is smaller than the diameter of the bore.

_____ **5** With one hand, hold the head end of the micrometer squarely against the cylinder wall (Fig. 5-7). Then, with your other hand, turn the thimble to increase the length of the micrometer. As you do this, feel for the maximum diameter by moving the rod end slightly from left to right and up and down. Read the mike when no left-to-right movement of the rod end is possible, and a slight drag is felt as you move the rod end up and down.

_____ **6** Write the cylinder diameter in the first space below. Then measure four other cylinders, and enter the measurements.

MEASUREMENTS WITH AN INSIDE MICROMETER

Piece 1	_____ in
Piece 2	_____ in
Piece 3	_____ in
Piece 4	_____ in
Piece 5	_____ in

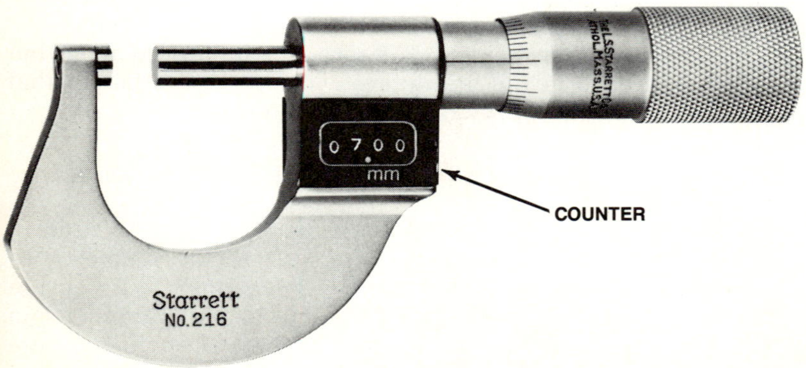

Fig. 5-8. Digital micrometer that reads in the metric system. (*The L. S. Starrett Company*)

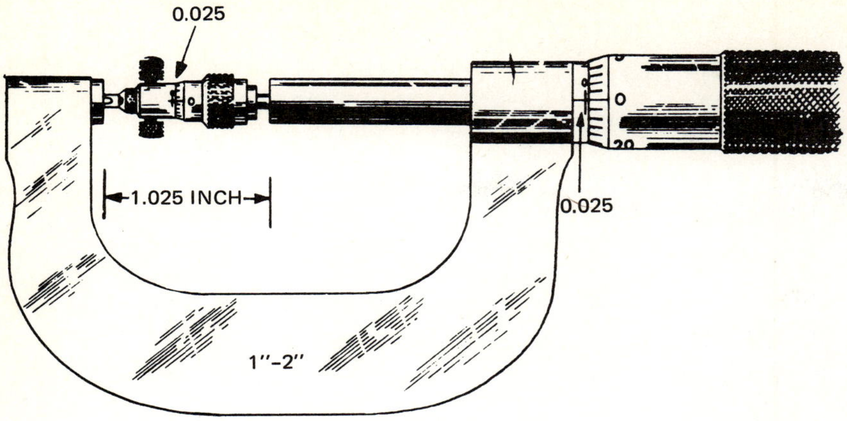

Fig. 5-9. Measuring the setting of an inside micrometer with an outside micrometer.

_____ 7 To check the accuracy of the inside micrometer, mike the inside micrometer (Fig. 5-9) with an outside micrometer. This transfers the reading from the inside mike to the outside mike. Both readings should be the same. If they are not, check the zero setting on each micrometer.

True-or-False Test — Job 5

1 The inside micrometer is used to measure shaft diameters. **1** _____

2 The micrometer hub markings are 0.025 in apart. **2** _____

3 To measure the diameter of a shaft, the micrometer should be clamped firmly to it. **3** _____

4 The spindle moves in or out as the thimble is turned. **4** _____

5 The thimble markings represent 0.001 in each. **5** _____

6 The markings on the hub and thimble of an inside micrometer are similar to those on the hub and thimble of an outside micrometer. **6** _____

7 Carrying a micrometer about in the pocket is apt to damage the micrometer, since dirt may get into the screw threads. **7** _____

8 Screw threads under the anvil cause the spindle to move in and out as the thimble is turned. **8** _____

9 When a shaft measuring 0.656 in is miked, the 21 mark on the thimble aligns with the line on the hub. **9** _____

10 When a shaft measuring 0.238 in is miked, the 13 mark on the thimble aligns with the line on the hub. **10** _____

Written Checkup — Job 5

1 Write down the procedure for measuring a shaft diameter with a micrometer. Explain how to read the micrometer.

2 Write down the procedure for measuring a cylinder bore with a micrometer.

NAME DATE CLASS

Objective: To learn how to use dial indicators.

Equipment: Engine block, engine crankshaft, generator or starting-motor armature, pieces of round bar stock or shafts.

Tools: Dial indicator and V blocks.

Job 6

DIAL INDICATORS

PROCEDURE

Part One READING A DIAL INDICATOR

_____ 1 The dial indicator is a gauge that uses a dial face and a needle to register measurements (Fig. 6-1). The dial indicator has a movable contact arm. When the arm is moved, the needle rotates on the dial face to show the arm movements in thousandths of an inch.

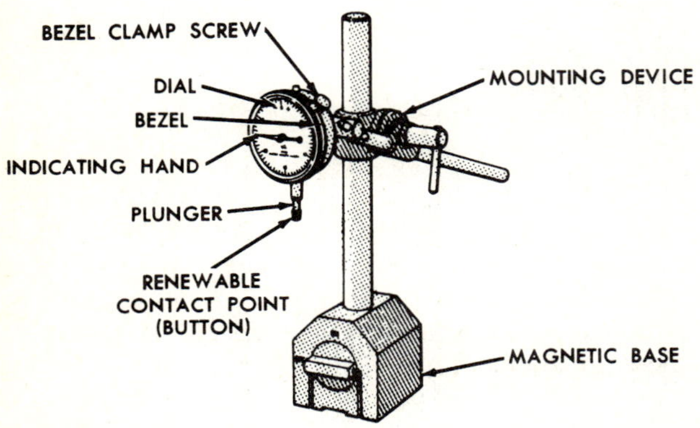

Fig. 6-1. Dial indicator. (*Ford Motor Company*)

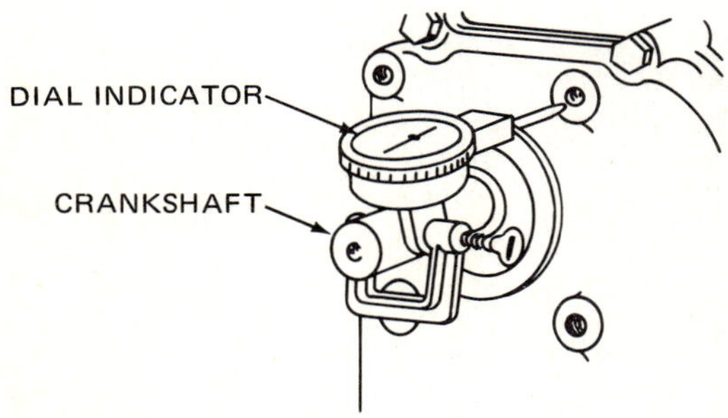

Fig. 6-2. Checking the crankshaft end play of a small engine with a dial indicator.

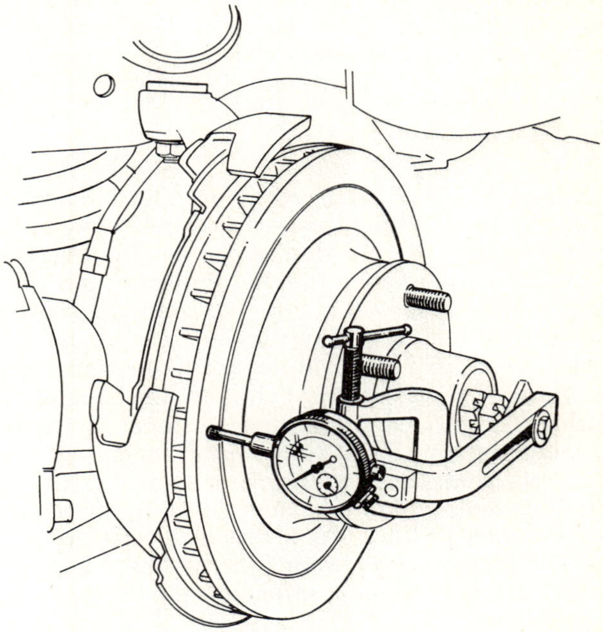

Fig. 6-3. Checking a disc-brake rotor for runout with a dial indicator. (*Buick Motor Division of General Motors Corporation*)

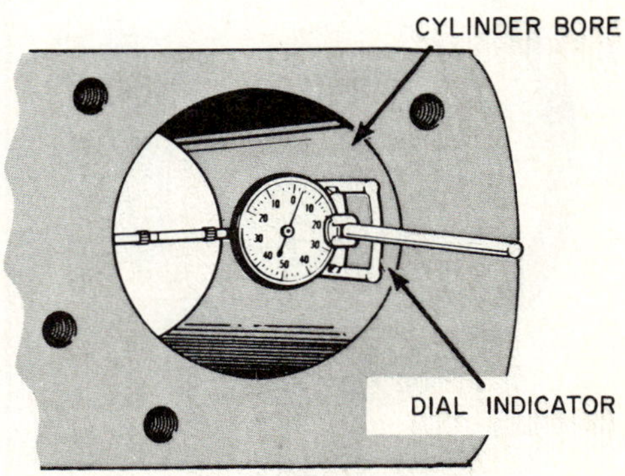

Fig. 6-4. Using a dial indicator to check for irregular wear of a cylinder bore.

_____ 2 The dial indicator is used to measure end play in shafts or gears (Fig. 6-2), movement of contact points, runout of disk brake rotors (Fig. 6-3), cylinder bores in engine blocks (Fig. 6-4), and so on. Figure 6-4 shows a dial indicator being used to measure the bore, or diameter, of a cylinder. As the dial indicator is moved up and down, any difference in the diameter causes the needle to move. Differences in the cylinder diameter at various points indicate cylinder wear. These special dial indicators for measuring cylinder wear are also called *cylinder-bore gauges*.

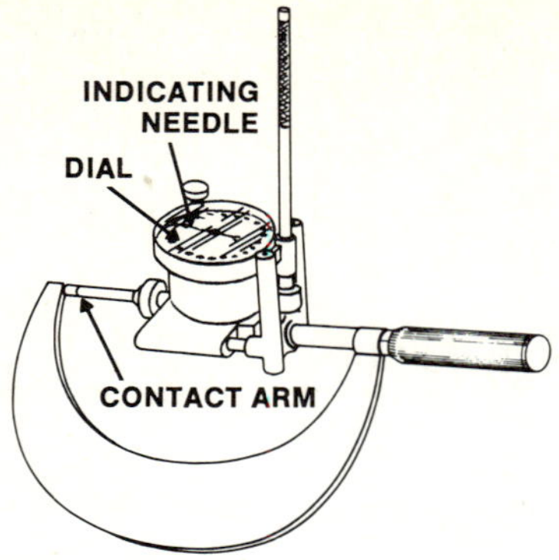

Fig. 6-5. A dial indicator can be used to measure the diameter, or bore, of an engine cylinder. Once the reading is taken, the dial is set to zero and the reading is measured with a micrometer. (*Pontiac Motor Division of General Motors Corporation*)

_____ **3** To use the dial indicator to find the actual diameter of a cylinder, insert the indicator in the cylinder. Note the position of the needle (Fig. 6-4).
_____ **4** Then remove the dial indicator and use a micrometer, as shown in Fig. 6-5. Some dial indicators can be locked before removal. On others, you must adjust the mike until the needle is in the same position as it was in the cylinder. Then, read the setting of the outside micrometer to get the actual cylinder bore, in thousandths of an inch.
_____ **5** There are two scales on some dial indicators (Fig. 6-6). The outer scale is usually marked in measurements of 1/1,000 of an inch (0.001 in). The inner scale records or counts the number of revolutions made by the large indicator needle.

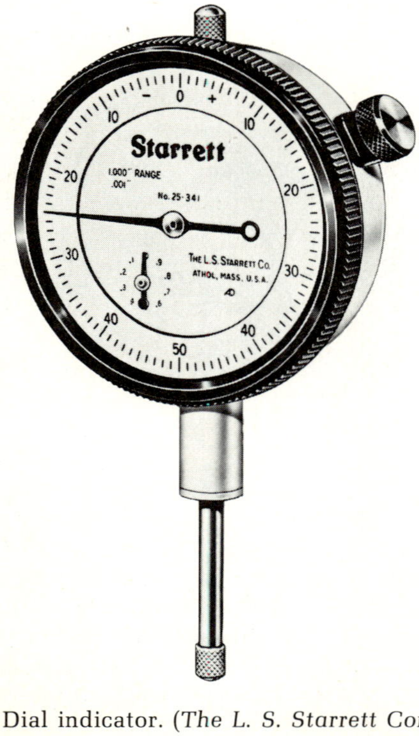

Fig. 6-6. Dial indicator. (*The L. S. Starrett Company*)

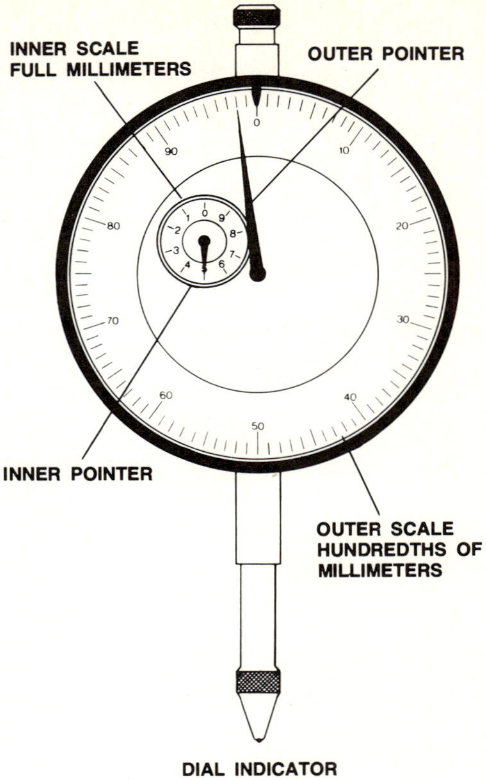

Fig. 6-7. Metric dial indicator. (*Volkswagen of America, Inc.*)

_____ **6** On the metric indicator (Fig. 6-7), the outer scale is divided into 100 divisions. Each division represents $\frac{1}{100}$ of a millimeter (0.01 mm). The inner scale is divided into 10 divisions. Each division is 1 mm. So, every complete revolution of the outer scale represents 1 mm.
_____ **7** Now, let's read the measurement shown on the dial indicator in Fig. 6-7. The needle on the inner scale has passed 4 but has not quite reached 5, making a reading of 4 mm.
_____ **8** To find out how much past 4 we've gone, we must add the reading on the outer dial. The dial on the outer scale determines the fractional (hundredths of a millimeter) reading. In Fig. 6-7 the outer needle is pointing to 98, which is read and written as 0.98 mm. Added together, the total reading is 4.98 mm.

Part Two USING A DIAL INDICATOR

_____ **1** Examine a dial indicator. As you can see, there is a movable contact button on one side. Press the button lightly with your finger. The needle on the dial fact turns as the button is moved. A ratchet-and-gear arrangement produces this motion.

NOTE: Do not repeatedly push on the contact button just to see the needle spin around. The internal mechanism is subject to damage or wear if the indicator is treated roughly.

Fig. 6-8. Using a dial indicator and V blocks to check the runout of an armature commutator.

_____ **2** Place the armature from a generator or starting motor in V blocks set up on a workbench or table. Set up the dial indicator, with the contact button resting on the commutator of the armature (Fig. 6-8).

_____ **3** Rotate the armature through one complete revolution. Read the out-of-round of the commutator on the dial indicator. Enter the total amount of out-of-round in the first space below. Then place several other shaft-type objects in the V blocks, and measure the out-of-round of each. Write each measurement below.

MEASURING SHAFTS WITH A DIAL INDICATOR

Shaft 1	_____ in	_____ mm
Shaft 2	_____ in	_____ mm
Shaft 3	_____ in	_____ mm
Shaft 4	_____ in	_____ mm
Shaft 5	_____ in	_____ mm

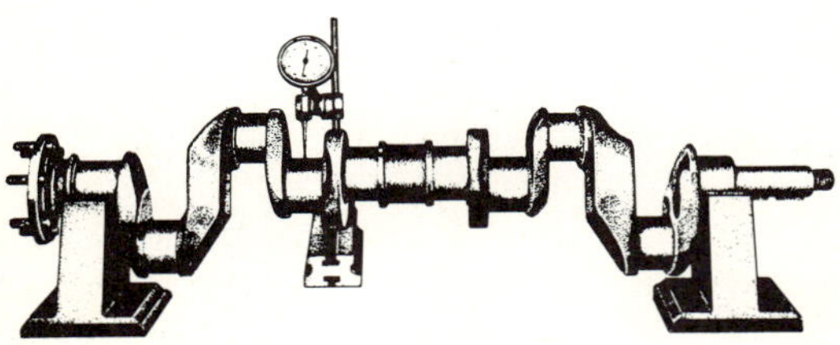

Fig. 6-9. Checking a crankshaft for alignment with V blocks and a dial indicator.

_____ **4** Now, set up a dial indicator to determine the runout, or out-of-round, of a crankshaft (Fig. 6-9). Place the crankshaft in V blocks. Rotate the crankshaft while the button of the dial indicator is held against a main bearing journal (Fig. 6-8).

_____ **5** In the spaces below, write the amount of out-of-round of each crankshaft journal. This is read, for each journal, as the total movement of the needle on the dial indicator.

MEASUREMENTS WITH A DIAL INDICATOR

Journal 1	_____ in	_____ mm
Journal 2	_____ in	_____ mm
Journal 3	_____ in	_____ mm
Journal 4	_____ in	_____ mm
Journal 5	_____ in	_____ mm

_____ **6** Dial indicators are delicate measuring instruments. They must be treated with great care. They must not be dropped or banged about. After they are used, they should be wiped clean and returned to their cases.

Multiple-Choice Test — Job 6

1 A dial indicator is a gauge that registers measurements on a **A** hub and thimble **B** steel scale **C** dial face with a needle **D** caliper **1** _____

2 The dial indicator is used to measure **A** shaft diameters **B** shaft length **C** shaft out-of-round **D** weight **2** _____

3 A quick way to measure engine cylinder wear is to use a **A** feeler gauge **B** cylinder-bore gauge **C** caliper **D** steel scale **3** _____

4 On the metric dial indicator, each division of the outer scale is **A** 1 mm **B** 10 mm **C** 100 mm **D** 0.01 m **4** _____

5 When the needle swings from one side of the zero mark to the other side, you add the two readings to get the **A** length **B** width **C** total height **D** total out-of-round **5** _____

Written Checkup — Job 6

1 Write down the procedure for setting up a dial indicator.
2 Explain the purpose of the small needle on the inside of the face of some dial indicators.
3 How do you set the dial-indicator needle to zero?
4 What is the maximum measurement that can be read on the dial indicator you are using?
5 Write down the procedure for setting up and using a dial indicator to measure the runout of a wheel on a car.

NAME DATE CLASS

Objective: To learn how to use vernier calipers.
Equipment: Pieces of round bar stock and pipe.
Tools: Vernier caliper.

Job 7

VERNIER CALIPERS

PROCEDURE

Part One READING THE VERNIER CALIPER

_____ **1** The main disadvantage of ordinary calipers (Job 2) is that they do not give a direct reading of the measurement taken. To find the actual measurement, you must measure the caliper setting with a steel scale or rule.

_____ **2** In 1631, Pierre Vernier, a French mathematician, invented an accurate, direct-reading scale for use in making linear measurements. When Vernier's scale was combined with an ordinary caliper, the result was a measuring instrument called a vernier caliper.

_____ **3** Basically, the vernier caliper is made of two graduated steel rules (Fig. 7-1). One rule is fixed and is called the fixed rule or frame. The fixed jaw is attached to one end of the frame (Fig. 7-1). The second rule is movable and slides along the frame. One end of the sliding rule has the other measuring jaw attached (Fig. 7-1). This jaw is called the adjustable, or sliding, jaw.

_____ **4** To measure an outside diameter with a vernier caliper, place the object snugly between the jaws of the caliper (Fig. 7-2). The number of inches is read on the fixed scale on the frame. To this, add the number of tenths that are seen between the last inch reading and the 0 on the vernier scale. Then the number of 0.025-in marks seen between the last tenth reading and the 0 on the vernier scale are added. Finally, read the number of lines from 0 on the reverse scale to the point where the line on the vernier scale coincides exactly with a line on the fixed scale. Each of these lines represents 1/1,000 of an inch (0.001 in).

_____ **5** Figure 7-3 shows the use of a vernier caliper to measure an inside diameter. Some vernier calipers have slightly different scales on one side; they are used to measure inside diameters. Other vernier calipers have different 0 marks to be used when measuring inside diameters.

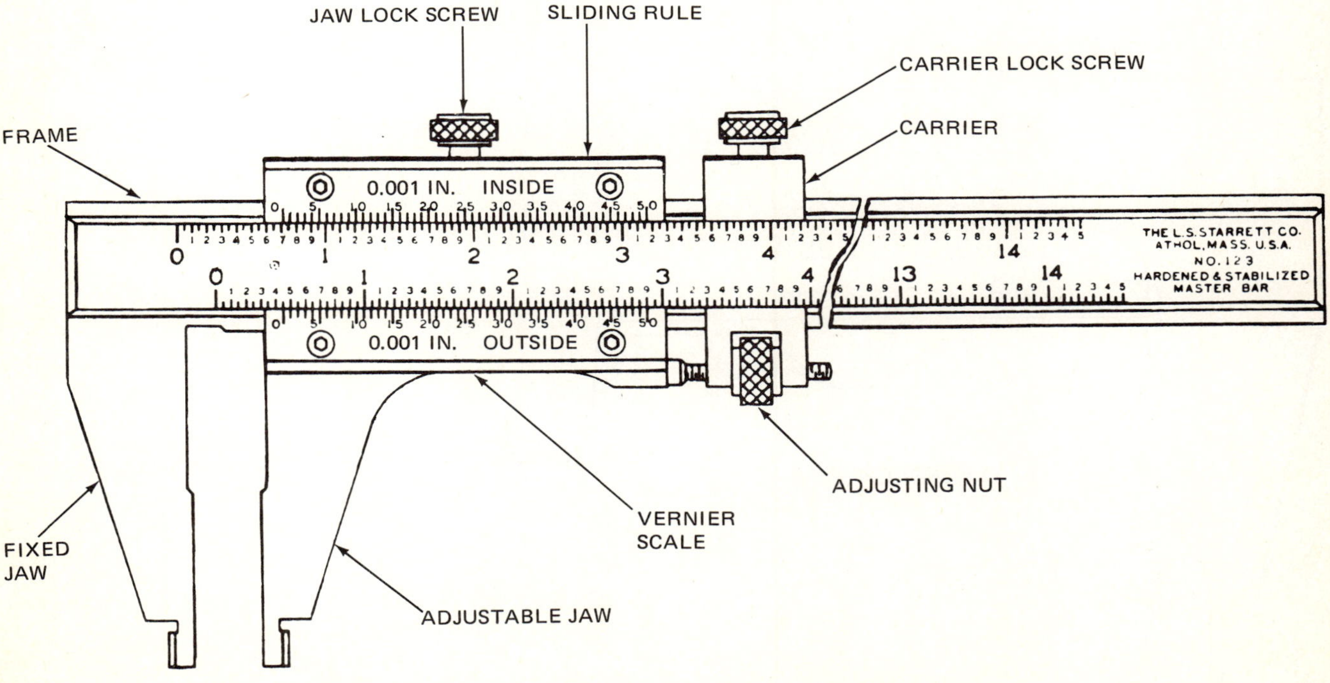

Fig. 7-1. Vernier caliper. (*The L. S. Starrett Company*)

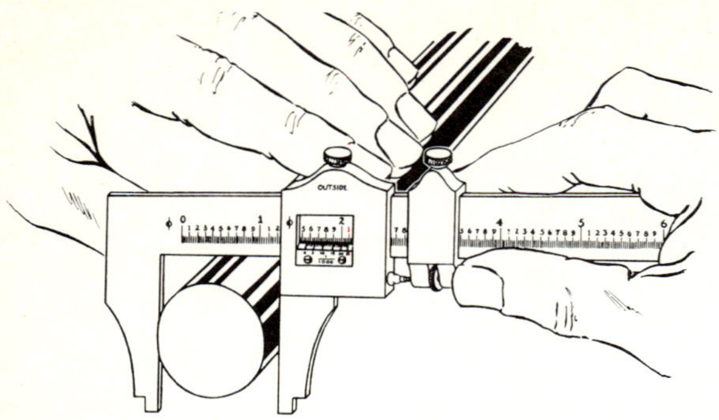

Fig. 7-2. Using the vernier caliper to measure an outside diameter.

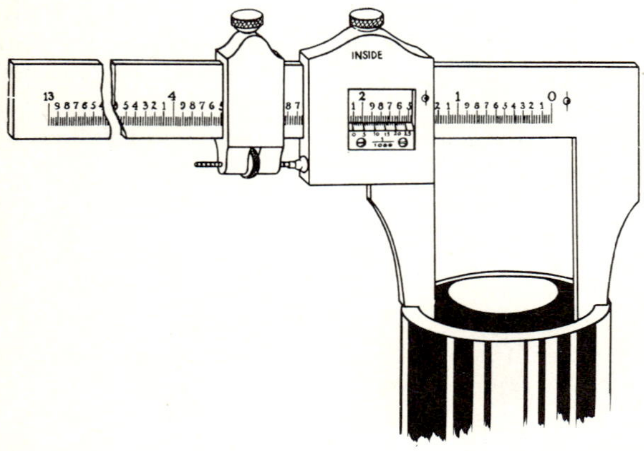

Fig. 7-3. The vernier caliper can also be used to measure an inside diameter.

_____ **6** Figure 7-4 shows a simplified metric vernier caliper that can be used for training purposes. On this caliper, the fixed scale is divided into 1-mm sections. The movable vernier scale is divided into 10 lines, each representing $\frac{1}{10}$ of a millimeter (0.10 mm).

_____ **7** To read the metric vernier caliper, count the number of millimeters (on the fixed scale) that precede the vernier-scale 0. Then, find a line on the vernier scale that coincides exactly with a line on the fixed scale. Count each line on the vernier scale from 0 to the two coinciding lines. Since each line is 0.10 mm, be sure to count by tens. This is the decimal portion of the reading. It must be added to the whole-millimeter reading to obtain the complete measurement.

_____ **8** A vernier caliper that has both metric and USCS scales has the metric scale on one side and the inch scale on the other side.

_____ **9** A dial caliper is shown in Fig. 7-5. With this precision measuring tool, outside and inside measurements can be read directly. Although dial calipers are relatively expensive, they are very quick and accurate to use. The dial caliper is read in about the same way as the dial indicator (Job 6).

Part Two USING A VERNIER CALIPER

_____ **1** Examine a vernier caliper (Fig. 7-1). Identify the fixed jaw and the adjustable jaw. Loosen the jaw lockscrew, and move the sliding rule. Then

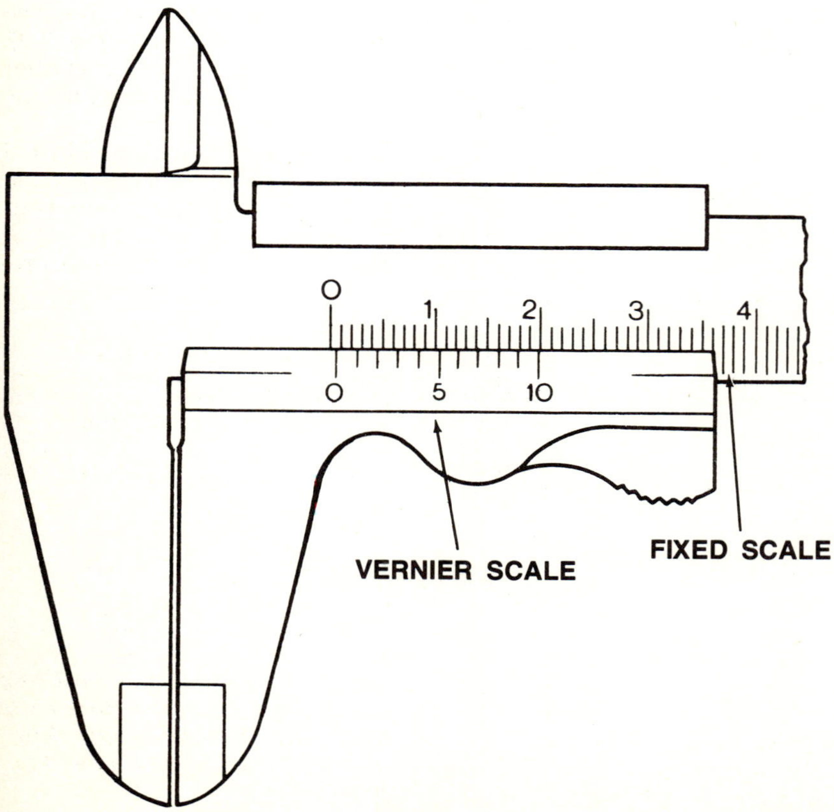

Fig. 7-4. Simplified metric vernier caliper. (*Volkswagen of America, Inc.*)

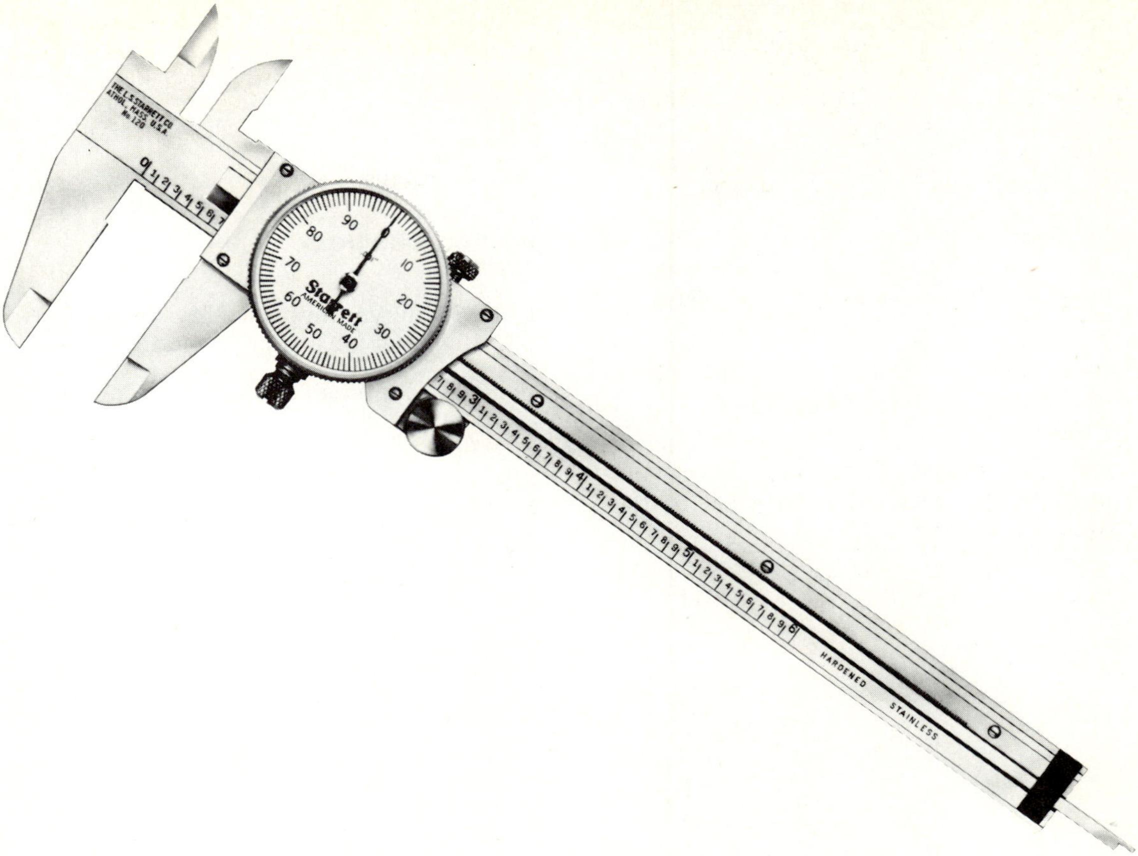

Fig. 7-5. Dial caliper. (*The L. S. Starrett Company*)

tighten the lockscrew, and read the measurement on the vernier scale. If you have any trouble reading the vernier scale, review Part One of this jobsheet.

_____ **2** Loosen both clamping screws on the vernier caliper, and open the jaws. Place a piece of round pipe or bar stock between the jaws, as shown in Fig. 7-2. Slide the movable jaw toward the piece being measured, until it is almost in contact. Then tighten the carrier lockscrew. This makes the adjusting nut operative.

_____ **3** Turn the adjusting nut until the adjustable jaw is in contact with the piece being measured and has the proper feel. Then read the measurement. Write your reading in the first space below. Measure the diameters of several different pieces of stock, and write the measurements in the spaces below.

MEASURING OUTSIDE DIAMETERS WITH A VERNIER CALIPER

Piece 1	_______ in	_______ mm
Piece 2	_______ in	_______ mm
Piece 3	_______ in	_______ mm
Piece 4	_______ in	_______ mm
Piece 5	_______ in	_______ mm

_____ **4** Now, let's use the vernier caliper to measure the inside diameter of a piece of pipe (Fig. 7-3). When measuring an inside diameter with a vernier caliper, remember to read the measurement on the scale marked "inside" (Fig. 7-3).

_____ **5** Place the jaws of the caliper inside the hole. Adjust the caliper until the proper feel is obtained. In the first space below, write down your measurement. Then measure the inside diameters of several different pieces of pipe. Write the measurements in the spaces below.

MEASURING INSIDE DIAMETERS WITH A VERNIER CALIPER

Piece 1	_______ in	_______ mm
Piece 2	_______ in	_______ mm
Piece 3	_______ in	_______ mm
Piece 4	_______ in	_______ mm
Piece 5	_______ in	_______ mm

_____ **6** Vernier calipers are precision measuring instruments. They must be treated with great care. They must not be dropped or banged about. After they are used, they should be wiped clean and returned to their cases.

Multiple-Choice Test — Job 7

1 A vernier caliper can be used to measure **A** inside diameters only **B** inside and outside diameters **C** outside diameters only **D** weight

1 ______

2 In measuring with a vernier caliper, the final adjustment is made with the **A** adjusting nut **B** jaw lockscrew **C** carrier lockscrew **D** frame

2 ______

3 A vernier caliper is basically **A** a micrometer **B** a dial indicator **C** two feeler gauges **D** two graduated steel rules 3 ______

4 Measurements on a vernier caliper are accurate to the nearest **A** 1 mm **B** 0.001 in **C** 0.025 in **D** 1 m (meter) 4 ______

5 A dual-dimensional vernier caliper has an inch scale on one side and, on the other side, the **A** micrometer scale **B** USCS scale **C** metric scale **D** English scale 5 ______

Written Checkup — Job 7

1 Write down the procedure for measuring the inside diameter of a pipe or cylinder with a vernier caliper.

2 How would you check a vernier caliper for accuracy?

3 Write down the procedure for measuring the outside diameter of a bar or shaft with a vernier caliper.

4 After using a vernier caliper in the shop, how do you properly clean and care for it?

Objective: To learn how to use depth, small-hole, and telescope gauges.

Equipment: Engine block, assorted stock with holes in it, overhead-valve type of cylinder head.

Tools: Depth, small-hole, and telescope gauges, micrometers.

Job 8

DEPTH, SMALL-HOLE, AND TELESCOPE GAUGES

PROCEDURE

Part One DEPTH GAUGE

_____ **1** The mechanic or service technician should know how to use many different types of measuring tools. Among these are the depth gauge, small-hole gauge, and telescope gauge.

_____ **2** Figure 8-1 shows three different types of depth gauge. The rule depth gauge is read like a steel rule (Job 2). The vernier depth gauge has greater accuracy than the rule depth gauge; it is read like a vernier caliper (Job 7). The micrometer depth gauge is more accurate than a vernier depth gauge; it is read like a micrometer (Job 5).

_____ **3** As you can see in Fig. 8-1, the depth gauge is used to measure depth, or the distance from a surface to a recessed part. The rule depth gauge is simply a sliding head on a steel rule (Job 2). The sliding head has two purposes. It is wide enough to bridge a hole or slot while allowing the rule to go into the recess. By bridging the slot, the sliding head holds the rule perpendicular to the surface being measured. This makes the measurement more accurate.

_____ **4** Many rule depth gauges have a measuring range of from 0 to 5 or 6 in [0 to about 125 or 155 mm]. Because they are so handy and so short, many technicians always carry one in their pocket.

_____ **5** Sliding heads for rule depth gauges are available in various sizes and thicknesses. However, regardless of size, almost all sliding heads have a clamp screw which permits the head to be clamped in any position. This locks the measurement on the depth gauge so it can be removed from the recess and read.

_____ **6** To measure the depth of a hole, you use the rule depth gauge as shown in Fig. 8-1. Loosen the clamp screw on the sliding head. Place the sliding head against the surface from which the depth is to be measured. Extend the steel rule into the hole. When the steel rule touches bottom, tighten the setscrew to maintain the setting. Then remove the rule depth gauge from the hole. Read the depth on the scale, at the edge of the sliding head that was against the upper surface.

_____ **7** The vernier depth gauge (center in Fig. 8-1) provides greater accuracy than the rule depth gauge. To use the vernier depth gauge, loosen the lockscrews. Slide the scale into the slot being measured until it almost touches the bottom. Then tighten the carrier lockscrew to make the adjusting nut operative.

_____ **8** Use the adjusting nut to extend the scale until the proper feel is obtained. Then lock the setting with the scale lockscrew. Remove the vernier depth gauge. Read the setting locked on the gauge. If you have trouble reading the vernier scale, refer to Job 7.

NOTE: The "proper feel" is obtained when you *first* notice or feel contact between the end of the scale and the bottom of the slot.

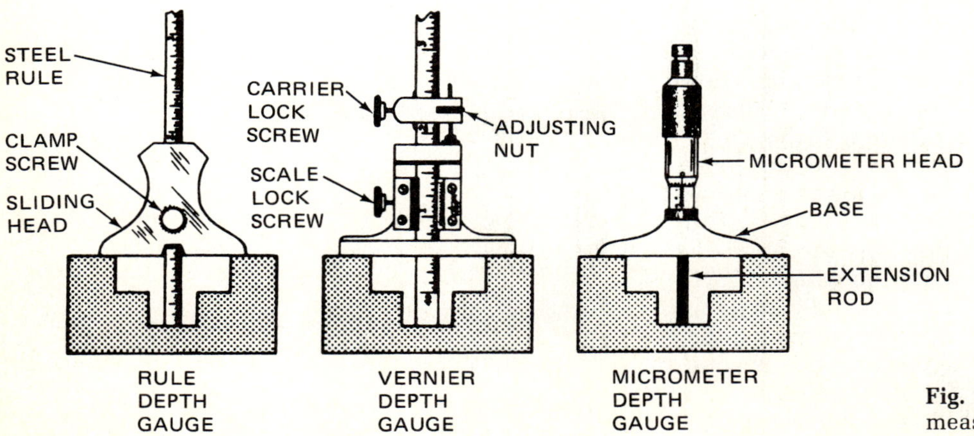

Fig. 8-1. Using depth gauges to measure the depths of holes.

_____ **9** The micrometer depth gauge (right in Fig. 8-1) has a flat base attached to a micrometer head. Extension rods of different lengths are available. They allow some micrometer depth gauges to have a measuring range of from 0 to 9 in [228.6 mm].

_____ **10** The micrometer depth gauge is used in the same manner as the micrometer (Job 5). The base of the micrometer depth gauge is placed on a surface, with the rod extending into the slot, hole, or recess. Then the thimble of the micrometer head is adjusted until the proper feel is obtained. Like micrometers, some micrometer depth gauges have a ratchet stop on the thimble.

_____ **11** When the proper feel is obtained, or the ratchet stop begins to slip, remove the micrometer depth gauge. Read the measurement on the micrometer. If you have any trouble reading the micrometer, refer to Job 5.

_____ **12** If an extension rod has been used to obtain the measurement, the length of the rod must be added to the reading on the micrometer. The total is the depth measurement.

_____ **13** Examine the depth gauge that you have. Use it to measure the depth of a hole or slot, as shown in Fig. 8-1. Then, measure the depths of several different holes. Write the measurements in the spaces below.

MEASURING WITH A DEPTH GAUGE

Hole 1	_______ in	_______ mm
Hole 2	_______ in	_______ mm
Hole 3	_______ in	_______ mm
Hole 4	_______ in	_______ mm
Hole 5	_______ in	_______ mm

Part Two SMALL-HOLE GAUGES

_____ **1** Small-hole gauges can be used to measure smaller holes or slots, that telescope gauges cannot fit into. Small-hole gauges are available in sets of four or more. They will measure distances from about $\frac{1}{8}$ to $\frac{1}{2}$ in [3.2 to 12.7 mm].

_____ **2** Figure 8-2 shows a small-hole gauge. As you can see it consists of a small split ball mounted on the end of a handle. The split ball expands when the knurled knob at the top of the handle is turned. Proper feel is obtained when you can feel a slight drag of the ball end against the sides of the hole. Then remove the gauge from the hole, and measure the size of the ball end with a micrometer. The micrometer reading is the diameter of the hole.

_____ **3** Examine a set of small-hole gauges. Turn the knurled knob at the top of a gauge, and watch the split ball move. Note the narrow shank to which the split ball ends are attached. The shank is not as wide as the ball ends. This permits the small-hole gauge to extend into a hole or small bore to measure the diameter some distance from the open end.

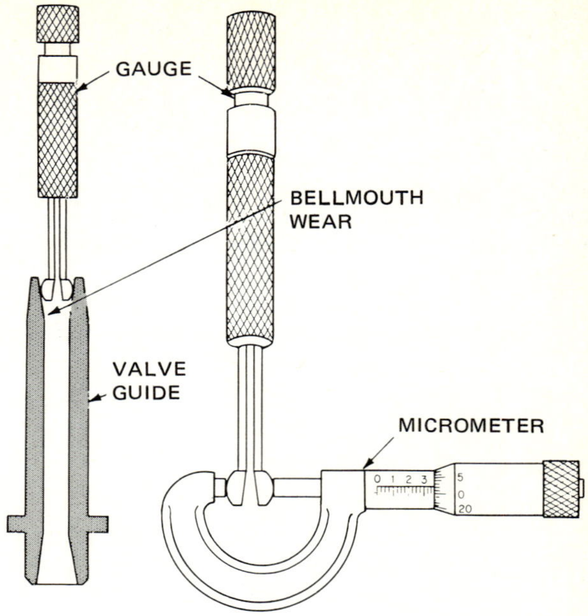

Fig. 8-2. Using a small-hole gauge to measure the size of a hole.

_____ **4** Obtain a 1-in micrometer and an engine cylinder head without the valves. Use the small-hole gauge and the micrometer to measure the diameter of each valve guide, as shown in Fig. 8-2.

_____ **5** Adjust the gauge so the ball is of minimum size. (Do this by turning the knurled tip of the handle.) Insert the ball into the hole, and turn the knurled tip to increase the size of the ball. Gradually increase the ball size until the ball drags slightly on the sides of the hole as you slide it in and out. Withdraw the ball, and measure its diameter with the micrometer (Fig. 8-2). Write the measurement for each valve guide in the spaces below.

MEASURING WITH A SMALL-HOLE GAUGE

Valve guide 1	_______ in	_______ mm
Valve guide 2	_______ in	_______ mm
Valve guide 3	_______ in	_______ mm
Valve guide 4	_______ in	_______ mm
Valve guide 5	_______ in	_______ mm
Valve guide 6	_______ in	_______ mm
Valve guide 7	_______ in	_______ mm
Valve guide 8	_______ in	_______ mm
Valve guide 9	_______ in	_______ mm
Valve guide 10	_______ in	_______ mm
Valve guide 11	_______ in	_______ mm
Valve guide 12	_______ in	_______ mm

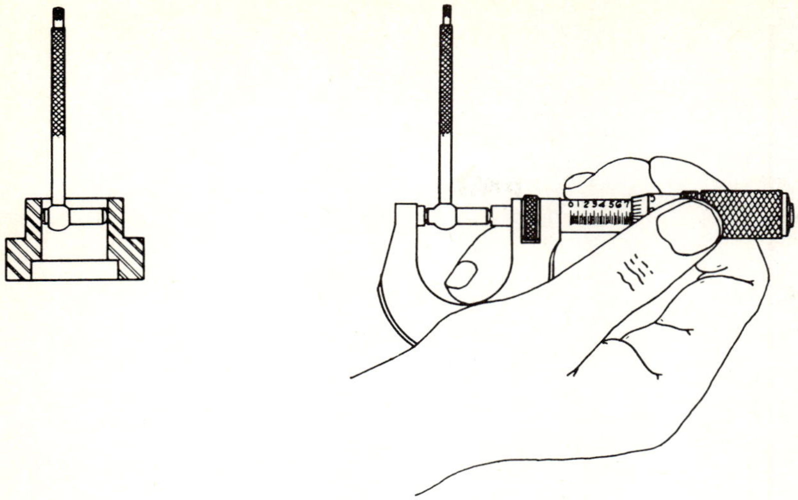

Fig. 8-3. Using a telescope gauge to measure the diameter of a cylinder. (*The L. S. Starrett Company*)

Part Three TELESCOPE GAUGES

_____ **1** Telescope gauges are used to measure hole diameters up to about 6 in [about 155 mm] (Fig. 8-3). A telescope gauge is a T-shaped measuring instrument. The shaft of the T is used as a handle, and the crossarm for measuring.

_____ **2** To use the telescope gauge, you loosen the knurled nut at the end of the gauge handle. Then compress the arms, with your hand, until their width is slightly smaller than the inside diameter to be measured. Tighten the nut to lock the arms in position.

_____ **3** Obtain an engine cylinder block. Insert the gauge into one cylinder, as shown in Fig. 8-3. Loosen the nut. A light spring inside the arms will push the adjustable end out, and the gauge will adjust to fit the cylinder. Make sure the telescoping end is at right angles to the cylinder wall. Then tighten the nut, and withdraw the gauge.

_____ **4** To find the bore of the cylinder, measure the telescope-gauge setting with a micrometer (Fig. 8-3). Write the cylinder diameter in the proper space below. Then use the telescope gauge and micrometer to measure each of the other cylinders. Write the measurements in the proper spaces below.

MEASURING WITH A TELESCOPE GAUGE

Cylinder No. 1	_____ in	_____ mm
Cylinder No. 2	_____ in	_____ mm
Cylinder No. 3	_____ in	_____ mm
Cylinder No. 4	_____ in	_____ mm
Cylinder No. 5	_____ in	_____ mm
Cylinder No. 6	_____ in	_____ mm
Cylinder No. 7	_____ in	_____ mm
Cylinder No. 8	_____ in	_____ mm

True-or-False Test — Job 8

1 The depth gauge is used to measure cylinder diameters. **1** _____

2 The small-hole gauge is used to measure the diameters of small holes. **2** _____

3 The telescope gauge is used to measure the diameters of cylinders. **3** _____

4 Turning the knurled tip of the small-hole gauge changes the diameter of the ball. **4** _____

5 Loosening the nut on the end of the telescope-gauge handle allows the spring to push the adjustable end out. **5** _____

Written Checkup — Job 8

1 Write down the procedure for using a small-hole gauge to measure the diameter of a hole.
2 Write down the procedure for using a telescope gauge to measure the diameter of a cylinder.
3 Write down the procedure for using a depth gauge to measure the depth of a hole or recess.

NAME DATE CLASS

Objective: To learn how to check clearances with shim stock and Plastigage.

Equipment: Various engines, partly disassembled so that the connecting-rod caps can be removed.

Tools: Torque wrench, shim stock, Plastigage, basic hand tools.

Job 9

MEASURING CLEARANCES WITH SHIM STOCK AND PLASTIGAGE

PROCEDURE

Part One CHECKING CLEARANCES WITH PLASTIGAGE

_____ **1** The precision-insert connecting-rod-bearing clearance is checked in the same manner as precision-insert main-bearing clearance. That is, either a piece of shim stock of the correct width or a strip of Plastigage is placed in the bearing cap, and the cap is reinstalled and tightened in place. First, let's discuss how to check the clearance using Plastigage.

_____ **2** When Plastigage is used, the oil must be drained, and the oil pan or cover removed. The bearing cap is then removed from the connecting rod that is to be checked.

_____ **3** If you were servicing the engine, you would thoroughly inspect the crankshaft bearing surface and the bearing itself. An examination of the bearing is necessary to determine the cause of wear or failure. Then you would make the corrections necessary to eliminate the cause of bearing failure and to avoid future failures from the same cause.

_____ **4** Examine the crankpin for roughness. You can use a micrometer (Job 5) to check it for taper and out-of-round. During engine service or overhaul, the crankshaft may be removed and reground or replaced.

_____ **5** The bearing clearance can be measured with the old bearing inserts before they are removed, or after new bearings are installed. It is essential, however, that the bearing clearance be checked every time new bearings are installed.

_____ **6** When using Plastigage, be sure all oil is removed from the crankpin and from the bearing surface. The best position of the crankshaft is with the crankpin about 30° (degrees) before bottom dead center. This allows you to measure the point of greatest wear. It also keeps the Plastigage away from the oil hole in the crankshaft. Do not turn the crankshaft with the Plastigage in place.

_____ **7** To check the bearing clearance with Plastigage, proceed as follows. Place a strip of Plastigage lengthwise in the center of the bearing cap (Fig. 9-1). Replace the cap. Use a torque wrench to tighten the cap nuts to the specified tension.

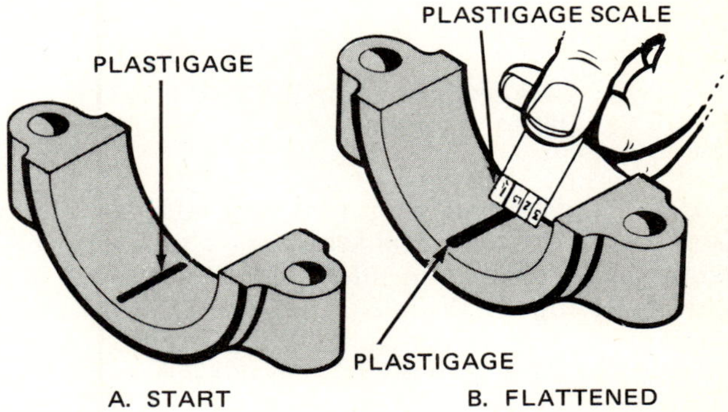

Fig. 9-1. Using Plastigage to check bearing clearance. (*A*) Lay a strip of Plastigage on the bearing in the rod cap. (*B*) After installing the cap and then removing it, measure the amount the Plastigage has flattened to determine the clearance.

_____ **8** Remove the cap. The flattened Plastigage will stick to either the bearing or the crankshaft. Do not touch or try to remove the flattened Plastigage until after measuring is completed.

_____ **9** Without removing the Plastigage, measure its compressed width at the widest point. Figure 9-1 shows this measurement being made. The width of the flattened Plastigage is being compared with the scale printed on the Plastigage envelope.

_____ **10** The scale printed on the Plastigage envelope is marked in thousandths of an inch and in millimeters. The wider and flatter the Plastigage (that is, the more it is flattened), the less the bearing clearance. The amount of flattening indicates the amount of clearance.

_____ **11** Uneven flattening indicates a tapered or worn crankpin or bearing. An excessively worn or tapered crankpin requires removal of the crankshaft for service.

If the Plastigage indicates that new bearings are required, you would select bearings of a size that gives the specified clearance. After installing them, you would measure the clearance again with Plastigage, to assure a proper fit.

_____ **12** Write down, in the first space below, the clearance measurement you obtained with the Plastigage. Then, use Plastigage to check the clearances of four more connecting-rod bearings or main bearings. Enter these measurements, in both inches and millimeters, in the spaces below.

MEASURING CLEARANCES WITH PLASTIGAGE

Bearing 1	______ in	______ mm
Bearing 2	______ in	______ mm
Bearing 3	______ in	______ mm
Bearing 4	______ in	______ mm
Bearing 5	______ in	______ mm

Part Two CHECKING CLEARANCES WITH SHIM STOCK

_____ **1** Before checking a bearing clearance with shim stock, turn the crankshaft, or shake the connecting rod from side to side. This allows you to feel how easily the crankshaft or connecting rod can be moved.

_____ **2** Obtain or make several pieces of shim stock. Each piece should be ½ in [12.7 mm] wide and 0.001 in [0.03 mm] thick.

_____ **3** Remove the cap from the bearing to be checked. Lightly lubricate the bearing, crankshaft, and shim stock with oil. Then place one piece of shim stock across the bearing (Fig. 9-2).

_____ **4** Install the cap. Use a torque wrench to tighten the cap nuts lightly. Then try to turn the crankshaft slightly. If no drag is felt, tighten all the nuts a slight additional amount.

_____ **5** Continue to tighten the nuts and check for crankshaft drag until the nuts have been torqued to the specified tension. Then turn the crankshaft not more than one-quarter turn in each direction. A slight drag should be felt, which indicates that the clearance is satisfactory. If it is, remove the cap. Take out the shim, and retorque the nuts in place.

_____ **6** If the crankshaft does not tighten up, the clearance is greater than the thickness of the shim stock.

_____ **7** When no drag is felt, remove the cap, and add another piece of shim stock. Repeat this procedure until a slight drag is felt while you turn the crankshaft. Then remove the cap, take out the shims, and retorque the nuts. The clearance is the thickness of the shims. If you are uncertain about the thickness of the shims, measure them with a micrometer (Job 5).

_____ **8** When a slight drag is felt, the bearing clearance is the same as the thickness of the shim stock in the bearing. If the crankshaft locks, the bearing clearance is less than the thickness of the shim stock in the bearing.

_____ **9** Write down, in the first space below, the clearance measurement that you obtained with shim stock. Then check the clearance of four more connecting-rod bearings or main bearings with shim stock. Enter these measurements in the spaces below.

MEASURING CLEARANCES WITH SHIM STOCK

Bearing 1	______ in	______ mm
Bearing 2	______ in	______ mm
Bearing 3	______ in	______ mm
Bearing 4	______ in	______ mm
Bearing 5	______ in	______ mm

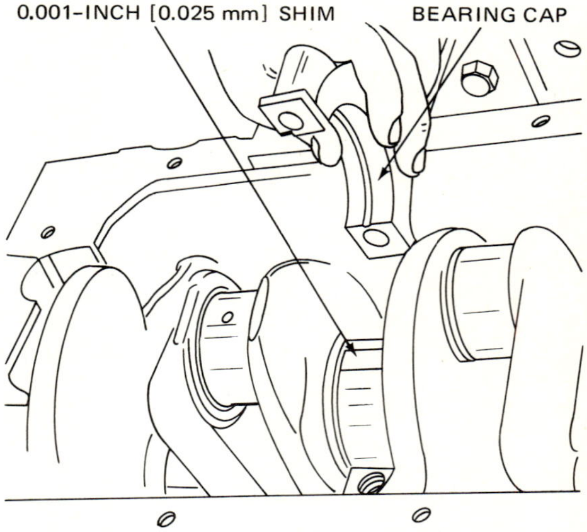

Fig. 9-2. Checking main-bearing clearance with shim stock. (*Chrysler Corporation*)

Multiple-Choice Test — Job 9

1 To check precision-insert connecting-rod bearings, you can use Plastigage or **A** a micrometer **B** a spring scale **C** shim stock **D** a rod gauge **1** _____

2 The more the Plastigage is flattened, the **A** thicker the bearing cap **B** less the bearing clearance **C** thinner the bearing cap **D** greater the bearing clearance **2** _____

3 If the connecting rod cannot be moved after a piece of shim stock has been placed in the bearing cap and the cap tightened, then the bearing clearance is **A** just right **B** excessive **C** more than the thickness of the shim stock **D** less than the thickness of the shim stock **3** _____

4 When Plastigage is used to measure bearing clearance, you note the **A** color of the Plastigage **B** drag on the rod **C** lengthening of the Plastigage **D** flattening of the Plastigage **4** _____

5 The thickness of shims required to cause a slight drag can be measured with **A** a feeler gauge **B** Plastigage **C** a micrometer **D** a steel scale

5 ______

Written Checkup — Job 9

1 Describe the procedure for using Plastigage to check bearing clearances.

2 Explain how the flattening of the Plastigage tells you what the bearing clearance is.

3 Describe the procedure for using shim stock to check bearing clearances.

4 With the shim stock installed, the crankshaft may (1) turn freely, (2) have a slight drag, (3) have a noticeable drag, or (4) lock up. Explain what the bearing clearance is in each of these four cases.

5 How do you obtain the clearance reading in millimeters when using Plastigage?

NAME DATE CLASS

Objective: To learn about screws, bolts, nuts, lockwashers, and cotter pins.

Equipment: Different kinds of screws, bolts, nuts, lockwashers, and cotter pins.

Tools: Steel rule, caliper or micrometer, thread gauge.

Job 10

AUTOMOTIVE FASTENERS

PROCEDURE

_____ **1** Fasteners are the parts that hold the engine and the automobile together. "Fastener" is the name given to any device that holds other parts together. Examples are the screw, the nut and bolt, and the stud and nut (Fig. 10-1). All are widely used in the automobile. Other types of fasteners include keys, snap rings, splines, and rivets. We discuss these in Job 11.

_____ **2** There are also permanent ways of fastening parts together, by welding and soldering. The car frame is welded together. Welding and soldering are specialized jobs often done in the auto body shop.

_____ **3** First, let us note the differences among screws, bolts, and studs. Machine screws (Fig. 10-1) are fasteners which are driven, or turned, into drilled and threaded holes in metal parts. A screw is turned down into a threaded hole to hold another part in place. There are many types of screws.

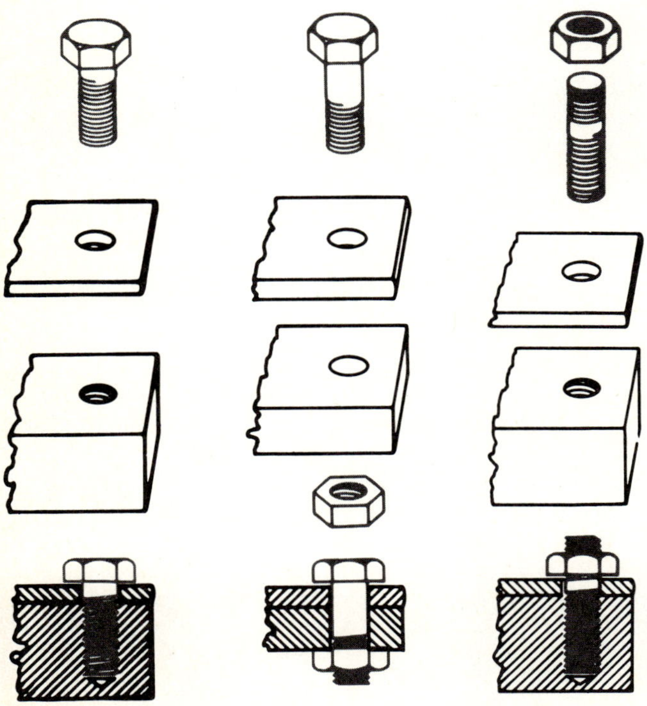

Fig. 10-1. Screw, bolt, and stud. Top shows the attaching parts separated, but aligned for assembly. Bottom shows the parts together, in sectional view.

_____ **4** Bolts require nuts (Fig. 10-1). The bolt is put through holes in parts to be attached to each other. Then a nut is turned onto threads on the bolt, to hold the parts together. Bolts and screws are the same. If they are used with nuts, they are called bolts. If they are screwed into a threaded part, they are called screws. However, the cylinder head is fastened to the cylinder block without the use of nuts. It is general shop practice to refer to these particular fasteners as "head bolts."

_____ **5** Studs (Fig. 10-1) have screw threads at both ends. A stud is screwed down into a threaded hole. Then a part to be held in place is installed over the stud. A nut is then turned down on the other end of the stud to hold the part in place.

_____ **6** Screws, bolts, and studs have external threads. Nuts and threaded holes for screws, bolts, and studs have internal threads. Threads are made with taps and dies (Jobs 24 and 25). Threads (or *screw threads,* to use the full name) are designated, or described in several ways. They are designated by size (by outside diameter or by a number); by the number of threads per inch (or *pitch*); by thread series (coarseness or fineness of thread); and by thread class (quality of finish and fit). See Table 10-1.

_____ **7** Many different bolts and screws are used in automobiles and other machines. Most bolts have hexagonal (six-sided) heads. Screws may have hexagonal (or "hex") heads, but they also have other provisions for driving. Wrenches and screwdrivers to drive the various types of bolts and screws are shown in Fig. 10-2.

_____ **8** Bolts and hex-head screws are made of materials of different strengths. The table in Fig. 10-3 shows the head markings that tell the quality of the bolt or screw. The minimum tensile strength is the pull, in pounds, that a round rod with a cross section of 1 in^2 [6.45 cm^2] can stand before it tears apart or breaks. Stronger screws are more expensive; they are used only where added strength is necessary.

_____ **9** The metric system of measurements is described in Job 3. Metric bolts, screws, and nuts are measured in millimeters. The threads are also different from the threads used on cars built in the United States. An automotive mechanic working on both domestic and foreign cars needs two sets of fasteners and two sets of wrenches.

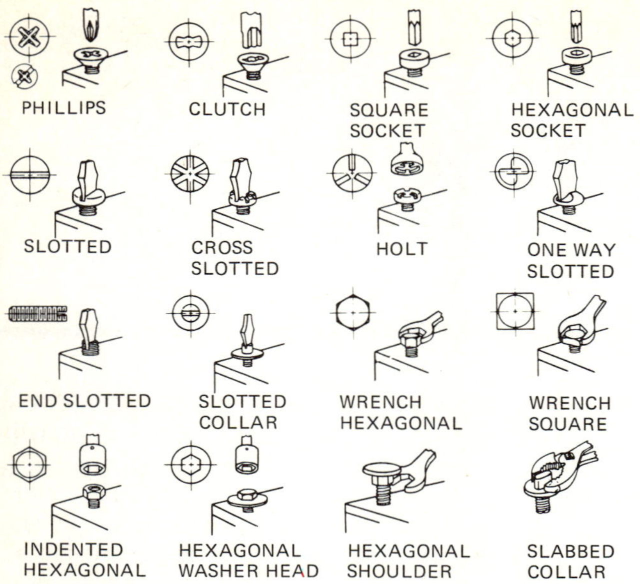

Fig. 10-2. Screwdrivers and wrenches required to drive various types of screws and bolts.

______ **10** Nuts come in various shapes (Fig. 10-4). Slotted and castle (or *castellated*) nuts are locked with a cotter pin. Other nuts are locked with lockwashers. Cotter pins and lockwashers prevent the nuts from working loose and dropping off. Lockwashers are also used under the heads of screws to keep them from loosening.

______ **11** Cotter pins (Fig. 10-5) are used with slotted and castle nuts. The bolt has a hole big enough for the cotter pin. First the nut is tightened, and the nut slots are lined up with the bolt hole. Then the cotter pin is inserted, and the two legs of the pin are bent as shown.

______ **12** Lockwashers (Fig. 10-6) are placed between the nut or screw head and a flat washer. The edges at the split (in the plain lockwasher) cut into the nut or screwhead. This keeps it from turning and loosening. Toothed lockwashers have many edges, to improve the locking effect. Sometimes the flat washer is not used. The lockwasher is placed between the nut or screw head and the machine part.

______ **13** Examine different screws and bolts. Note the types of heads on them. Select several bolts of different types and sizes. With a steel rule, caliper or micrometer, and thread gauge (Fig. 10-7), take various measurements as noted in Fig. 10-8. In the table on p. 46, record the measurements in the spaces provided. The threads per inch can be counted or checked with a thread gauge (Fig. 10-7).

______ **14** Examine various nuts. Select a nut to fit a particular bolt by using the thread gauge.

______ **15** Examine the various types of lockwashers (Fig. 10-6). Note that some lockwashers depend, for locking, on split or twisted sections biting into the bolt or nut and machine part. Others depend, for locking, on tangs that are bent up along the flat on the bolt head or nut.

______ **16** A still different locking action is obtained with a cotter pin (Fig. 10-5). Examine some cotter pins. Note that, on some applications such as wheel-bearing locknuts, the cotter pin is inserted through slots in the nut and a hole in the shaft. Note how the cotter-pin ends are cut off to length and then bent back around the nut to lock it in place.

USAGES IN VEHICLES	SOME USED	MUCH USED	FOR SPECIAL EQUIPMENT	CRITICAL POINTS	COMPETITION MAXIMUM REQUIREMENTS
TYPICAL APPLICATIONS	FENDERS	BELL HOUSINGS	HEAD BOLTS	BEARING CAPS	RACE CARS
MINIMUM TENSILE STRENGTH P.S.I.	64,000	105,000	133,000	150,000	160,000
MATERIAL	LOW-CARBON STEEL	MEDIUM-CARBON STEEL	MEDIUM-CARBON STEEL	ALLOY STEEL	SPECIAL ALLOY STEEL
QUALITY	INDETERMINATE	MINIMUM COMMERCIAL	MEDIUM COMMERCIAL	BEST COMMERCIAL	BEST QUALITY
HEAD MARKINGS	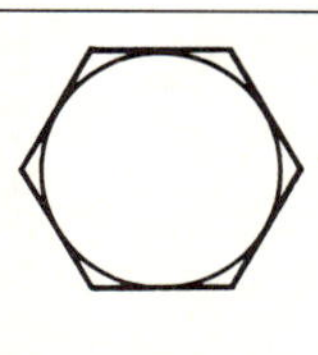	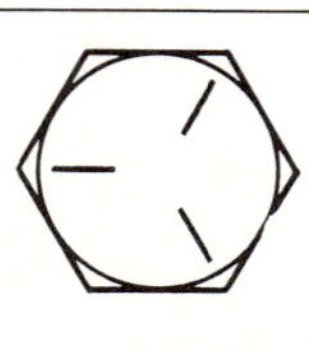	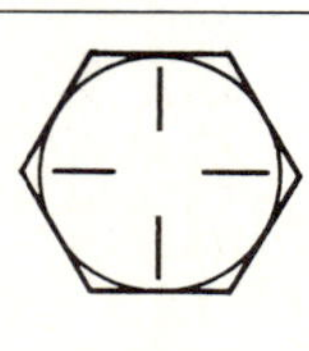	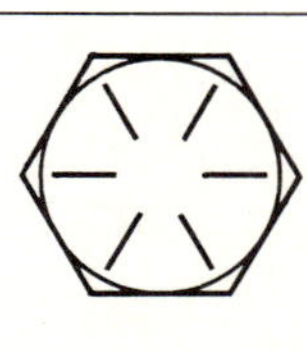	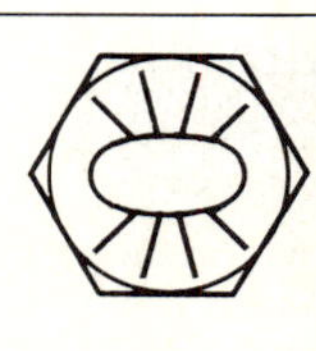

Fig. 10-3. Meaning of the bolt head markings.

TABLE 10-1

SIZE	DIAMETER (DECIMAL)	THREADS PER INCH COARSE (UNC OR NC)	FINE (UNF OR NF)	EXTRA FINE (UNEF OR NEF)
0	0.0600	. . .	80	
1	0.0730	64	72	
2	0.0860	56	64	
3	0.0990	48	56	
4	0.1120	40	48	
5	0.1250	40	44	
6	0.1380	32	40	
8	0.1640	32	36	
10	0.1900	24	32	
12	0.2160	24	28	32
1/4	0.2500	20	28	32
5/16	0.3125	18	24	32
3/8	0.3750	16	24	32
7/16	0.4375	14	20	28
1/2	0.5000	13	20	28
9/16	0.5625	12	18	24
5/8	0.6250	11	18	24
3/4	0.7500	10	16	20
7/8	0.8750	9	14	20
1	1.0000	8	12	20
1 1/8	1.1250	7	12	18
1 1/4	1.2500	7	12	18
1 3/8	1.3750	6	12	18
1 1/2	1.5000	6	12	18

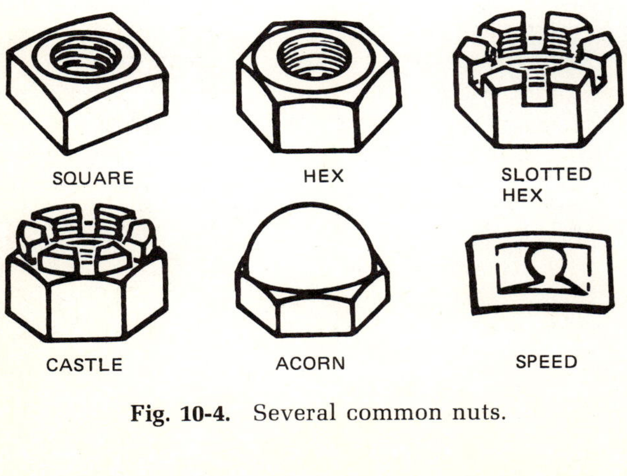

Fig. 10-4. Several common nuts.

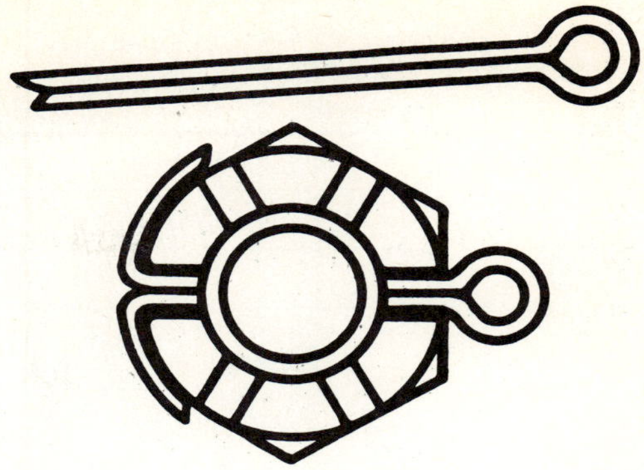

Fig. 10-5. Cotter pin before installation (top), and after installation through the hole in the bolt and the slot in the nut.

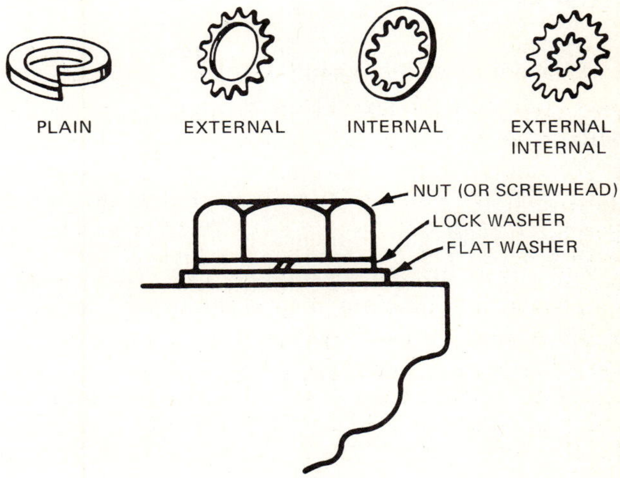

Fig. 10-6. Top, lockwashers. Bottom, a plain lockwasher installed between a flat washer and a nut or bolt head.

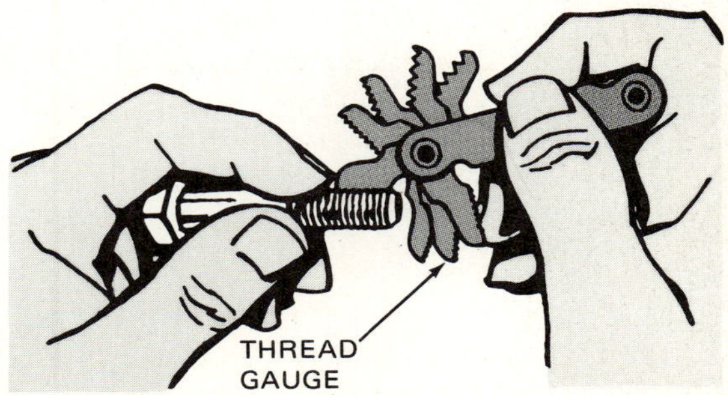

Fig. 10-7. Using a thread gauge.

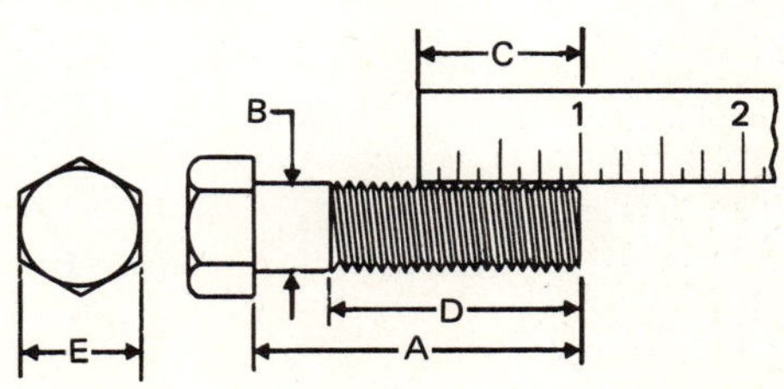

Fig. 10-8. A hex-head bolt with its parts named: *A*, length; *B*, diameter; *C*, pitch, or threads per inch; *D*, length of thread; *E*, size of wrench required to fit head.

BOLT NUMBER	A LENGTH	B DIAMETER	C THREADS PER INCH	D LENGTH OF THREAD	E WRENCH SIZE
______	______	______	______	______	______
______	______	______	______	______	______
______	______	______	______	______	______
______	______	______	______	______	______
______	______	______	______	______	______
______	______	______	______	______	______

True-or-False Test — Job 10

1 The width of a hexagonal bolt head, across two flats, indicates the size of the wrench to use on the bolt. **1** ______

2 The cotter pin is inserted through a hole in the shaft or bolt. **2** ______

3 The thread gauge measures the length of the threads in inches. **3** ______

4 Threads per inch may be measured with a rule or checked with a thread gauge. **4** ______

5 The wrench size required on a bolt is proportional to the bolt length. **5** ______

6 The locking action of many lockwashers depends on the sharp edges of the washer, which bite into the bolt head or nut to hold it in position. **6** ______

7 The legs of a cotter pin must be bent in opposite directions around a nut. **7** ______

Written Checkup — Job 10

Make a list of various fastening and locking devices used in automobiles. Write down several uses of each. Then add the names of the tools most often used to tighten or remove each of the fastening and locking devices. For example, you might write

Bolt: attaches cylinder head, carburetor; tightened with wrench or socket.

NAME DATE CLASS

Objective: To learn about snap rings, keys, splines, and rivets, and how to install them.

Equipment: Assorted snap rings, keys, splined devices, rivets and flat stock with drilled holes.

Tools: Pop rivet gun, ball-peen hammer, Truarc snap-ring pliers.

Job 11

ASSEMBLING DEVICES

PROCEDURE

Part One SNAP RINGS

_____ **1** There are two general types of snap rings. *External* snap or retaining rings (Fig. 11-1) prevent end-to-end movement of a gear or collar on a shaft. *Internal* snap or retaining rings (Fig. 11-1) are used in machine parts to keep shafts or other components in position.

_____ **2** The external snap ring must be expanded, or stretched, with special snap-ring pliers (Fig. 11-2). Then it is slipped over the shaft and into an undercut on the shaft. The internal snap ring must be squeezed so it can slip into the hole and into an undercut (Fig. 11-2).

_____ **3** The Truarc retaining ring is a special type of snap ring. Its two lips have holes for the pin ends of the special snap-ring pliers (Fig. 11-2). Thus, there is less chance of the pliers slipping off the ring during removal or installation. These rings are made for both internal and external installation.

_____ **4** Snap rings are made of spring steel. They can cause severe injury if the proper tools are not used to remove and install them.

_____ **5** Not all snap rings must be removed and installed with special pliers. Figure 11-3 shows one type of snap ring which is installed with needle-nose pliers. In Fig. 11-3, the snap ring is called a "piston-pin lock."

Part Two KEYS AND SPLINES

_____ **1** Keys and splines are used to lock gears, pulleys, collars, and similar parts to shafts. Figure 11-4 shows a typical key installation. The key is a wedge-shaped piece of metal. It fits into slots (or *keyways*) in both the shaft and the part being installed on the shaft. The key thus locks the shaft and collar together.

_____ **2** This half-round type of key is also called a *Woodruff* key. It is strong enough to carry a heavy load when fitted and seated properly. The gears at the front of most engines are usually positioned and timed by a key.

_____ **3** Square keys, like Woodruff keys, can carry heavy loads. Square keys are used to prevent wheels, gears, cams, and pulleys from turning on a shaft.

_____ **4** Splines (Fig. 11-5) are external and internal teeth cut in the shaft and the installed part. When the pinion, gear, pulley, or collar is installed, it is the same as having many keys between it and the shaft. Often, splines fit loosely so that the installed part can move back and forth on the shaft. But the splines force the installed part to rotate with the shaft. Splines can be curved, as in Fig. 11-5, or straight.

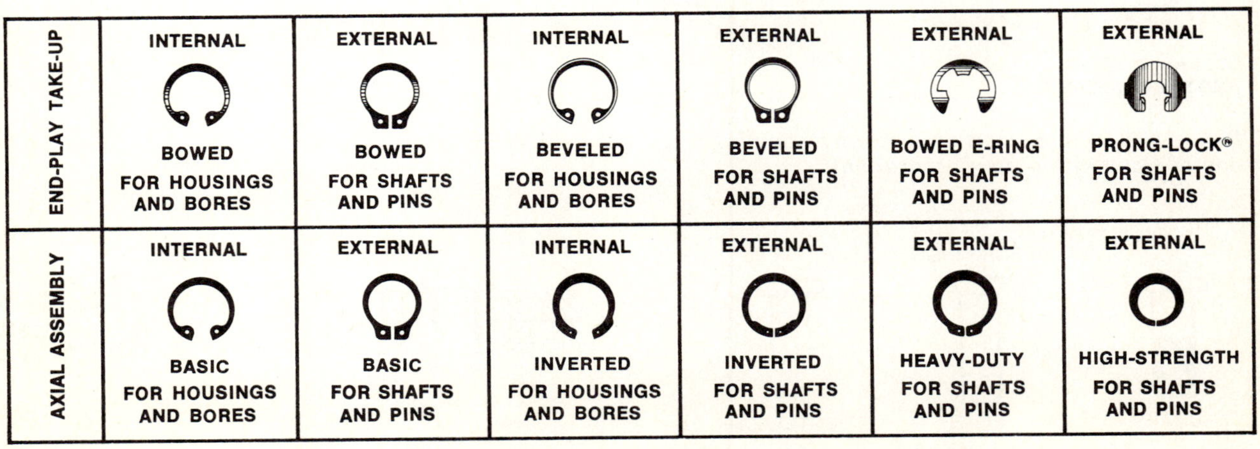

Fig. 11-1. Various types of internal and external Truarc retaining or snap rings. (*Waldes Kohinoor, Inc.*)

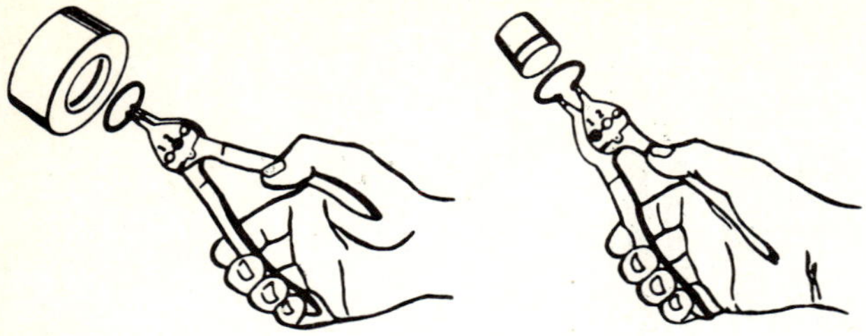

Fig. 11-2. Using Truarc retaining-ring pliers to install internal and external retaining rings. (*Waldes Kohinoor, Inc.*)

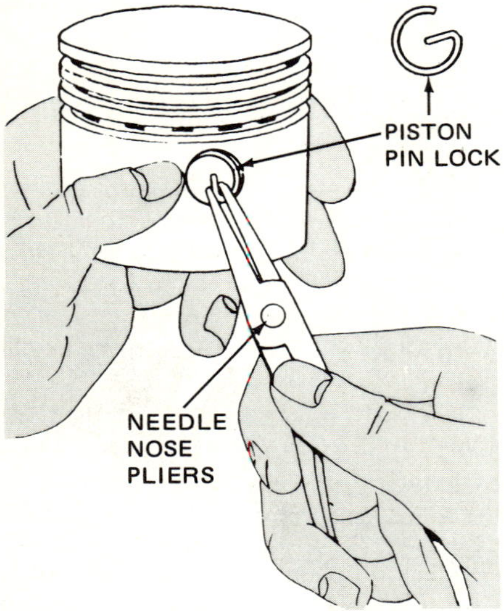

Fig. 11-3. Removing a piston-pin lock, or snap ring, with needle-nose pliers. (*Briggs & Stratton Corporation*)

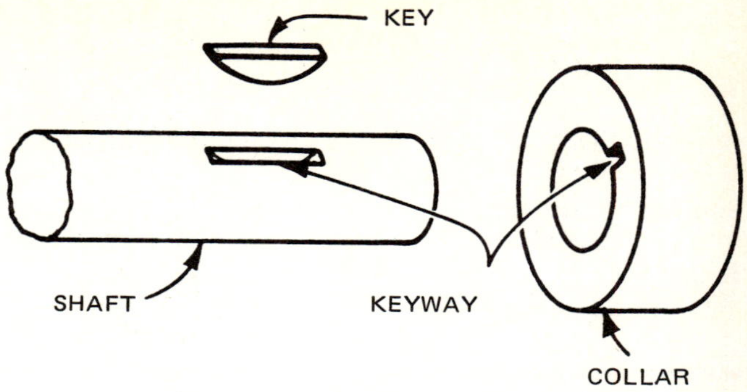

Fig. 11-4. Using a key.

Part Three RIVETS

_____ **1** Rivets (Fig. 11-6) are metal pins used to hold parts together more or less permanently. In the automobile, rivets hold the brake linings on some brake shoes. They are also used to keep the clutch lining in place. Rivets are used to hold the transmission linkage together, and to attach trim and upholstery to the car.

_____ **2** One end of the rivet has a head (Fig. 11-6). After the rivet is in place, a driver, or hammer-and-rivet set, is used to form a head on the other end. Rivets are usually removed by cutting off one end with a chisel. They can also be driven out with a punch, or drilled out.

_____ **3** In blind holes, where one end of the rivet cannot be reached for flattening, a Pop rivet can be used (Fig. 11-7).

_____ **4** Now, practice installing rivets. Obtain two

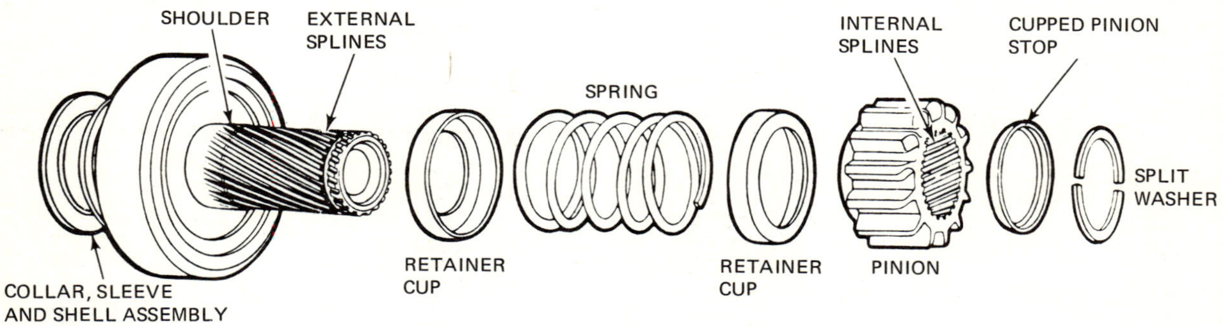

Fig. 11-5. Disassembled starting-motor overrunning-clutch drive, showing internal and external splines. (*Delco-Remy Division of General Motors Corporation*)

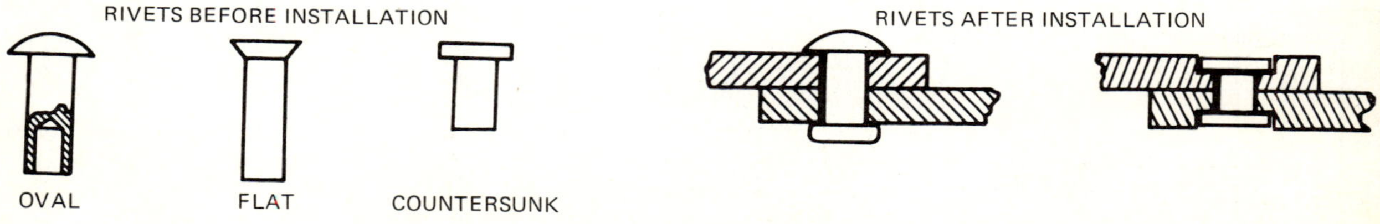

Fig. 11-6. Rivets before installation (left), and after installation (right).

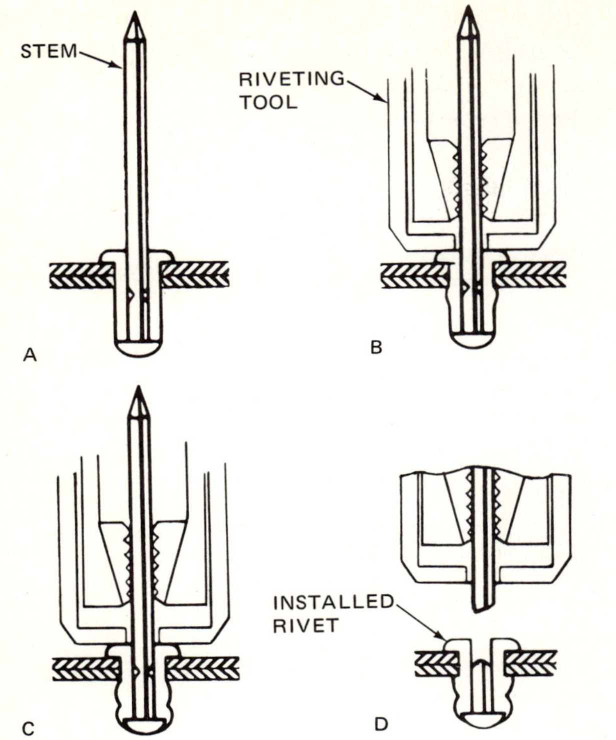

Fig. 11-7. Steps in installing a Pop rivet in a blind hole. (*A*) Rivet is placed in hole. (*B*) Riveting tool is placed over stem of rivet. (*C*) Rivet head compresses. (*D*) Stem "pops" or snaps off. (*Avdel Corporation*)

pieces of metal with holes in them. From the rivet assortment, select the proper size and length of rivet to fasten the two pieces of metal together. Figure 11-6 shows a few of the many types of rivets available.

_____ **5** Place the rivet through the holes so that both pieces of metal are properly positioned. With a hammer, peen over (or *upset*) the narrow end of the rivet (Fig. 11-8). To do this, you will need to place the head end of the rivet on the anvil part of the vise, or on some large, heavy metal work face. This prevents the hammer blows from driving the rivet out of the metal pieces being joined. It ensures that the rivet will draw the pieces into place. Solid-shank rivets may be peened with the peen end of a ball-peen hammer (Fig. 11-8).

_____ **6** Next, with the peen end of the hammer, flatten and round the head of the rivet (Fig. 11-9). For a more finished job, a rivet header tool can be used to round the head.

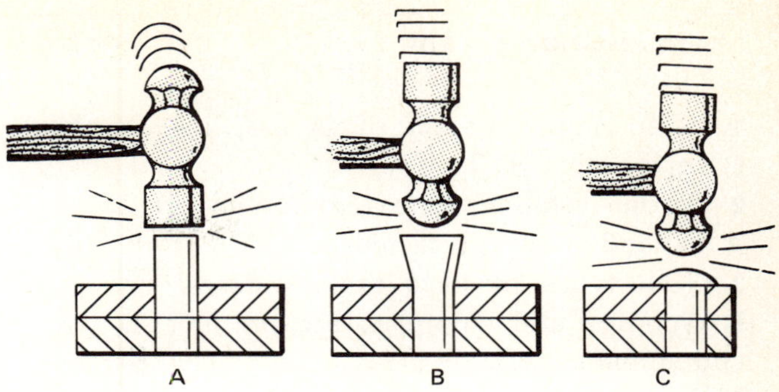

Fig. 11-8. Using a ball-peen hammer to install a rivet. (*Ford Motor Company*)

_____ **7** In some types of hollow-shank or tubular rivets, the rivet end must be rolled over with a special rivet set. When the rivet set is used, the end of the rivet is not driven down, but is rolled against the part being riveted.

_____ **8** Install at least five rivets. Your shop may have machines which you can use to rivet brake lining to a brake shoe, and friction lining to a metal clutch disc. If pop rivets are available, install a few of them also.

_____ **9** When you have installed at least five rivets, examine them closely. Compare each rivet you installed to Fig. 11-9, which shows examples of correct rivet installation, and common troubles. In the spaces below, write down the types of rivets that you installed. Then note whether they are correctly installed, or any installation troubles that you can identify.

INSTALLING RIVETS

RIVET	TYPE	INSTALLATION
1	_____	_____
2	_____	_____
3	_____	_____
4	_____	_____
5	_____	_____

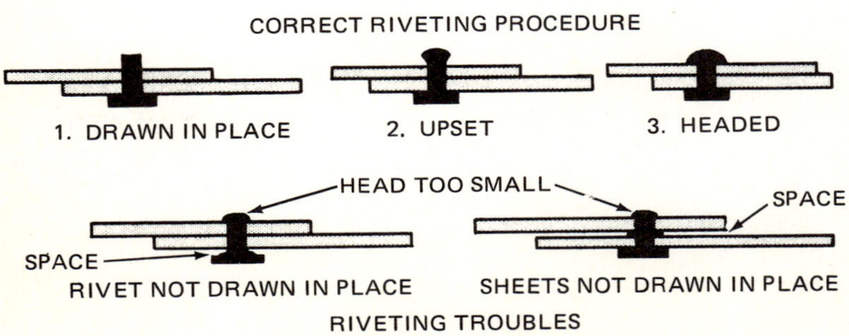

Fig. 11-9. Examples of correct and incorrect rivet installation.

Multiple-Choice Test — Job 11

1 Snap rings are made of **A** cast iron **B** plastic **C** spring steel **D** lead 1 ______

2 To be installed, an external snap ring must be **A** expanded **B** contracted 2 ______

3 Two types of keys are **A** dowel and taper **B** Woodruff and square **C** pin and belt **D** screws and studs 3 ______

4 In the automobile, rivets may be used to hold **A** brake linings on brake shoes **B** cylinder heads to the cylinder block **C** pistons to the connecting rods **D** transmission gears in place 4 ______

5 Rivets that can be installed in a blind hole by one person using a special rivet gun are called **A** hollow rivets **B** flat rivets **C** countersunk rivets **D** Pop rivets 5 ______

Written Checkup — Job 11

1 Explain how a rivet works.
2 What are the two different types of snap rings?
3 Explain how a key works, and why it is used.
4 Describe the action of internal and external splines.

NAME DATE CLASS

Objective: To learn how to solder and tape a splice.
Equipment: Pieces of wire.
Tools: Soldering iron or soldering gun.
Supplies: Cloths, solder, sal ammoniac, sandpaper.

Job 12

SOLDERING

Safety: *Handle the hot soldering iron with care to avoid getting burned. Do not twist or pull the electric cord. Never lay the hot iron on the cord; this will burn through the insulation and cause a short or expose the bare wire. The hot soldering iron, when not in use, should be placed on its stand.*

PROCEDURE

_____ **1** Soldering is a metal-joining, or fastening, process. A low-melting-point metal called *solder* is heated until it melts. Then it flows into the spaces between two parts, such as wires, that are to be joined together. When the solder cools and hardens, it forms a bond between the two parts.

_____ **2** Soldering is used to form good electrical connections between bare wires and between a bare wire and a terminal. Soldering can also be used to close the seams in layers or sections of sheet metal, and to hold parts together physically. However, soldered joints do not support loads well. Thus, welding is preferred for load-carrying joints. If another means of fastening two load-carrying pieces is used (for example, riveting or bolting), then soldering can be used to seal the joints. Note that not all metals can be soldered, but the metals used in electric wiring can be soldered.

_____ **3** To solder, you need three things: solder, flux, and a heat source. For automotive service work, an electric soldering iron or soldering gun (Fig. 12-1) is usually most satisfactory. Today, most technicians use the soldering gun because it reaches operating temperature quickly and cools rapidly.

_____ **4** The purpose of the soldering gun is to apply heat to the wire, terminal, or splice. Then, when solder is applied, it melts and runs into the spaces between the wires. When the splice cools, the solder hardens so that the connection is permanently protected.

_____ **5** Figure 12-1 shows a cutaway view of a soldering gun. Note that it has one or two small spotlights which turn on with the gun, to illuminate the work area. The soldering tip is made of copper and is easily replaceable. Pressing the trigger switch with your finger turns the gun on. The tip reaches operating temperature in a few seconds.

_____ **6** Before a surface can be soldered, it must be clean. Chemicals, wire brushes, and sandpaper can be used to prepare surfaces for soldering. If you try to solder dirty surfaces, the solder will not flow into the fine cracks between the surfaces.

_____ **7** Solder is made of a mixture, or *alloy,* of lead and tin. It is available in bars and as ribbon, solid wire, or hollow wire filled with flux. For general automotive service work, rolls of flux-filled solder wire are very convenient.

_____ **8** Flux is required for soldering. It is usually a paste or a liquid that cleans the surfaces to be soldered. Flux also prevents the formation of oxides during the heating process and before the solder starts to flow. Oxides forming on the surface of a hot

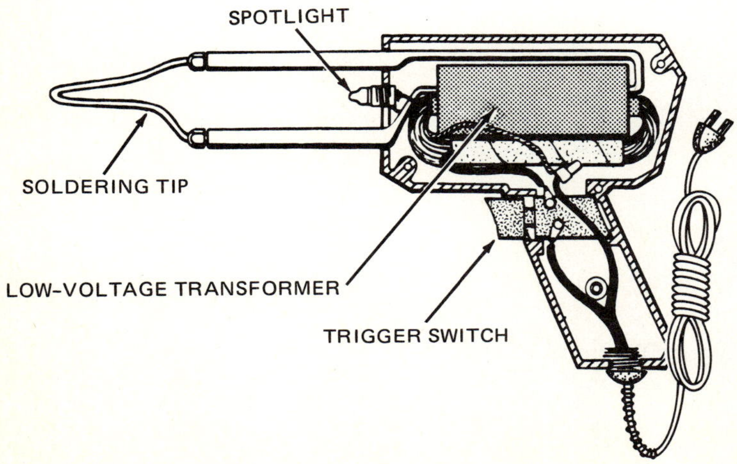

Fig. 12-1. Cutaway view of an electric soldering gun. (*Ford Motor Company*)

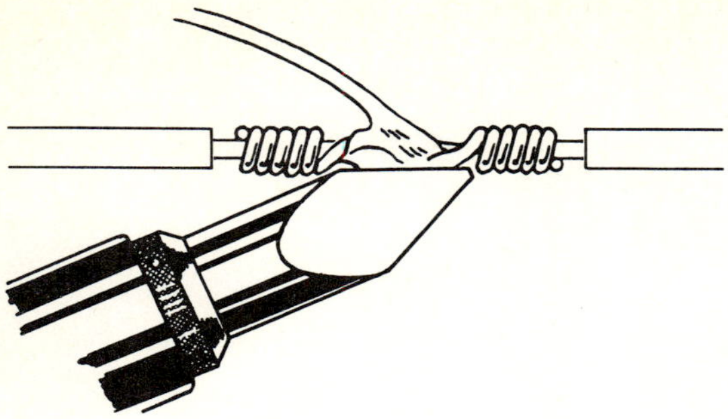

Fig. 12-2. Soldering an electrical connection.

copper wire might prevent good electrical connection.

_____ **9** The proper flux for electrical work is *rosin flux*. Acid fluxes must not be used in electrical work. They leave acid on the parts and cause the connections to deteriorate. Before using a roll of solder, always check the label to be sure that it reads "Rosin Core Solder."

_____ **10** Before a soldering iron or soldering gun is used, it must be clean, smooth, and tinned. If the tip is rough or pitted, its faces may be smoothed with a fine-cut file. The tip may be either hot or cold when it is filed. As the tip approaches soldering temperature, it will melt solder. The solder will spread over the tip, giving the tip a silvery (tinned) appearance. If flux-core solder wire is used, the tip need not be fluxed before the solder is applied. The flux in the core spreads on the tip, permitting the solder to spread.

_____ **11** Do not run a soldering gun too long and heat the tip too much. When the tip gets too hot, it changes in color from silver to yellow. This indicates that the solder "tinning" has oxidized. An overheated tip splatters any solder it touches. Such a tip will not solder well. It must be cooled, cleaned, and retinned before further soldering is attempted.

_____ **12** When soldering an electrical connection, hold the tip of the soldering iron against the parts to be soldered. Also hold the solder against the parts. In other words, don't melt the solder with the tip of the soldering iron. Instead, heat the parts to be soldered until they are hot enough to melt the solder. A good procedure to use is shown in Fig. 12-2.

_____ **13** One of the most important jobs for the automotive technician is soldering wire splices. To practice making splices, obtain a length of wire. Cut it into sections about 6 in [152.4 mm] long. Strip the insulation from at least 2 in [50.8 mm] at each end of two wires. (Insulation can be removed with a wire stripper, cutting pliers, or a knife. If a knife is used, hold it at an angle away from your body, as though you were using it to sharpen a pencil. Do not nick the wire with the knife, since this weakens the wire and may cause it to break.) Then scrape the enamel coating of insulation from the exposed ends of the wires.

_____ **14** On multiple-strand wire, (more properly called *cable*), use special care to avoid damaging or breaking off strands. Twist the wire strands together so that there are no stray strands sticking out.

_____ **15** To make a common splice, cross the wires and then twist them together, as shown in Fig. 12-3. After turning each wire over the other as shown in steps A, B, and C in Fig. 12-3, clip off the wire ends. Press the ends down with pliers so that they will not push through the insulating tape after it is put on.

_____ **16** Now solder the splice as described above, using rosin-core solder. When the solder has spread over the splice, remove the soldering iron. Hold the splice steady until the solder cools and hardens. Remember never to use acid flux on electrical connections.

_____ **17** Check the splice by comparing it to Fig. 12-2. Then have your instructor check it. If your splice is okay, make and solder four more splices. Then, using plastic electrical tape, tape one of the splices as shown in Fig. 12-4.

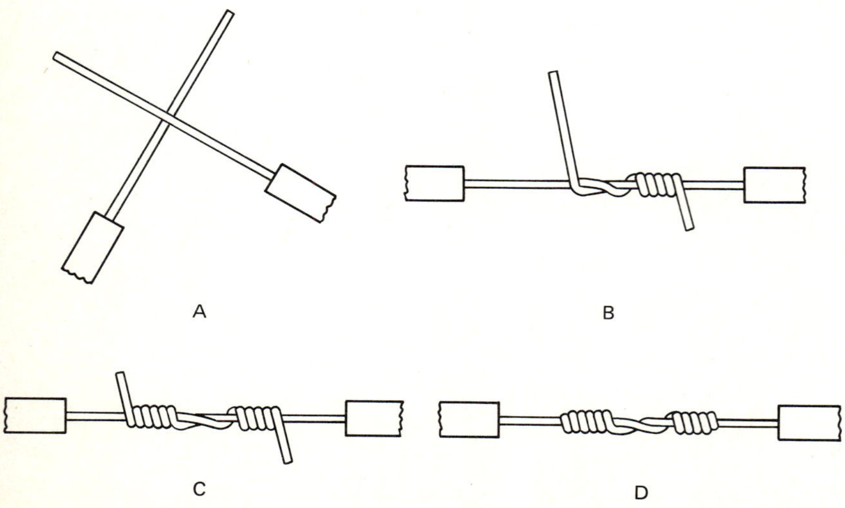

Fig. 12-3. Splicing two electric wires.

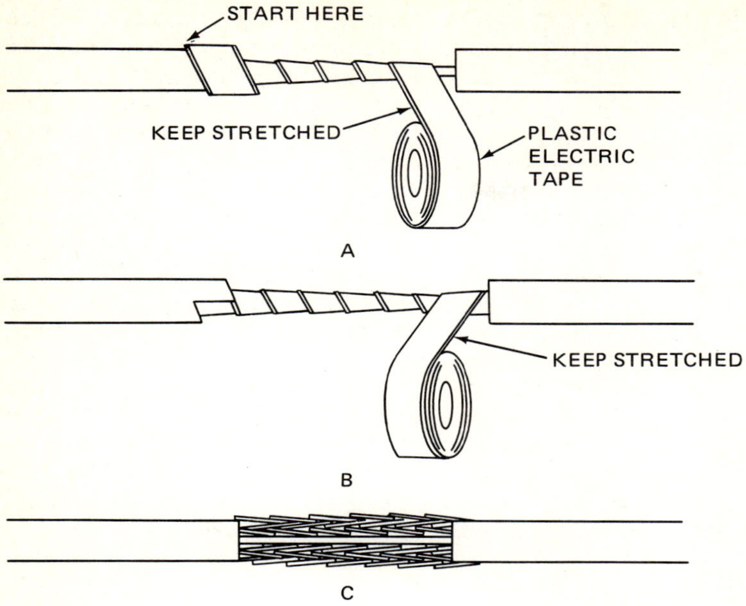

Fig. 12-4. Wrapping spliced electric wires with insulating tape.

True-or-False Test — Job 12

1 Surfaces to be joined by soldering must be clean. **1** ______

2 The tip of the soldering iron is made of iron. **2** ______

3 The soldering-iron tip should be tinned before it is used. **3** ______

4 Overheating the tip causes it to burn so that it must be cleaned and retinned. **4** ______

5 Rosin flux should always be used when electric connections are soldered. **5** ______

6 Flux cleans oxides away and prevents them from forming on surfaces to be joined by soldering. **6** ______

7 For soldering, the soldering-iron tip should be red hot. **7** ______

8 Electric connections should be made with acid flux. **8** ______

9 If the soldering tip is rough, it should be smoothed with a file. **9** ______

10 The solder should melt on the parts being soldered. **10** ______

Written Checkup — Job 12

1 Write down the procedure you followed in doing a soldering job.

2 Write down the procedure for taping a splice with plastic electrical tape.

NAME DATE CLASS

Objective: To learn about the types of screwdrivers and their uses.

Equipment: Various screws, blocks in which screws can be assembled and driven.

Tools: Various types of screwdrivers.

Job 13

SCREWDRIVERS

PROCEDURE

_____ **1** Most machines, including engines, are put together with bolts, screws, and nuts. To take an engine apart, or disassemble it, these bolts, screws, and nuts must be turned and taken off so that the parts can be separated. The most common tools used to perform this job are wrenches and screwdrivers. The screwdriver is used to drive, or turn, screws with slotted or recessed heads. The most common type is the driver for slotted screw heads (Fig. 13-1).

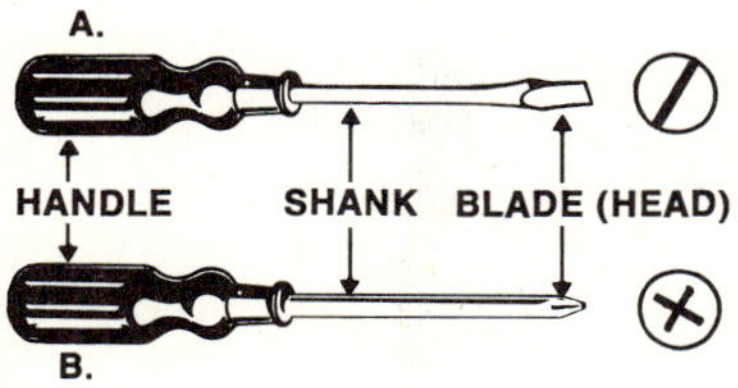

Fig. 13-1. (*A*) Typical slotted-head screwdriver. (*B*) Phillips-head screwdriver.

_____ **2** The typical screwdriver (Fig. 13-1) has a handle and a shank. The end of the shank is flattened to form a blade. The blade tip should be flat on the end so that it fits squarely into the screw slot (Fig. 13-2). If it is tapered, it will tend to rise out of the screw slot when it is turned. This makes the removal or installation of a screw more difficult. A worn screwdriver tip should be reground.

_____ **3** Some screws have what seem to be two slots at right angles. The Phillips screw is one of these. The Reed and Prince screw is another. Each requires a special screwdriver with a tip that fits the crossed slots (Fig. 13-3). The Phillips screw is widely used on automobile trim and molding. There is less chance that the screwdriver will slip out of the slots and damage the finish. Three sizes of Phillips-head screwdriver, 4 in, 6 in, and 8 in [101.6 mm, 152.4 mm, and 203.2 mm], are enough for most automotive work.

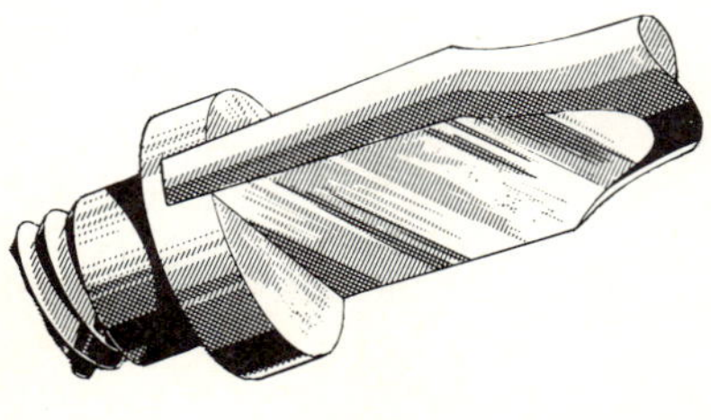

Fig. 13-2. The screwdriver tip should fit the screw slot properly.

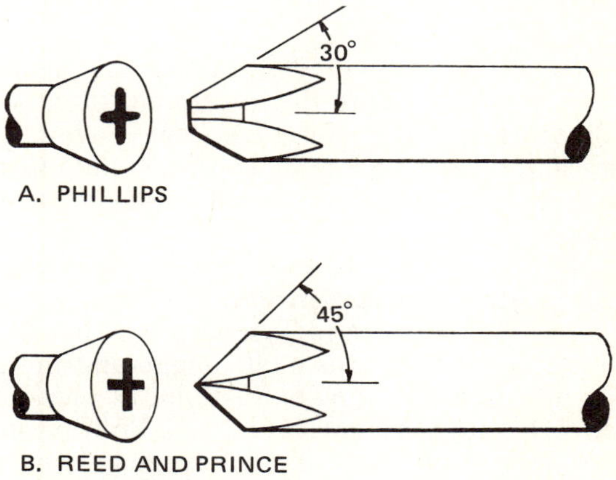

Fig. 13-3. Difference between the Phillips and the Reed and Prince screwdrivers.

_____ **4** The offset screwdriver is a variation of the standard screwdriver (Fig. 13-4). A screw that is set in an awkward place can be reached and turned with the offset screwdriver. The blade tips are set at right angles. First one end of the screwdriver can be used, and then the other, to keep the screw turning.

_____ **5** The Allen wrench (Fig. 13-5) is a special form of turning tool. It has a hexagonal (six-sided) shape that fits into a hexagonal hole in the head of a screw.

_____ **6** As you can see in Fig. 13-1, the part of the screwdriver that you grip is the handle. Extending from the handle is the shank. On the drive end of the shank is the blade. Most shanks are made of steel High-quality screwdrivers have hardened blades to prevent rapid wear.

_____ **7** Screwdrivers are classified by the length of the shank, from the handle to the drive end of the blade. Common screwdriver sizes range from $2\frac{1}{2}$ to 12 in [63.5 to 304.8 mm]. However, there are many special-purpose smaller and larger screwdrivers.

_____ **8** Screwdriver handles are made of wood. plastic, or metal. Nowadays, many general-purpose screwdriver handles are plastic. One advantage of a

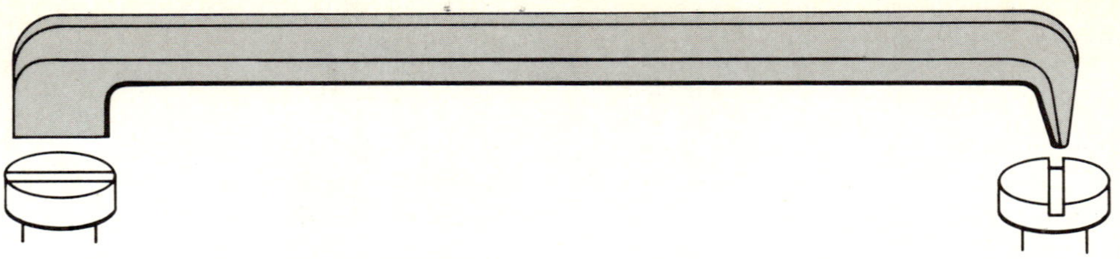

Fig. 13-4. Offset screwdriver.

plastic handle is that it insulates the user from most electric shocks. Also, the plastic will take some tapping or light hammering without splitting as wooden handles do.

_____ **9** In a screwdriver, the diameter of the shank is proportional to the width and thickness of the blade. For heavy-duty use, some shanks are square so that a wrench may be used to help turn the screwdriver. Pliers should never be used to turn a screwdriver.

_____ **10** Examine Fig. 13-3 closely. Note the difference in drive angle between the Phillips screwdriver and the Reed and Prince screwdriver. The Phillips screwdriver uses a 30° angle, and the Reed and Prince uses a 45° angle. Reed and Prince screwdrivers are not interchangeable with Phillips screwdrivers. If one of these screwdrivers is used on a screw of the other type, the tool or the screw head will be ruined.

_____ **11** The purpose of the four-way screw slot used in Phillips and in Reed and Prince screws is to prevent the screwdriver from slipping. Never try to turn a Phillips screw with a standard screwdriver. Both the tool and the screw head will be damaged.

_____ **12** Never use a screwdriver to check an electric circuit.

_____ **13** Never hold work in your hands while trying to turn a screw in it with a screwdriver. If the screwdriver slips, you may stab yourself seriously.

_____ **14** Never allow any part of your body to be in front of the screwdriver blade. This is a good safety rule to follow when using any pointed tool.

_____ **15** Examine various screwdrivers. Notice how the blade tips are designed to fit into the slot of the screw (Figs. 13-1 to 13-4). Note that standard screws have a single slot, while Phillips screws have a cross slot.

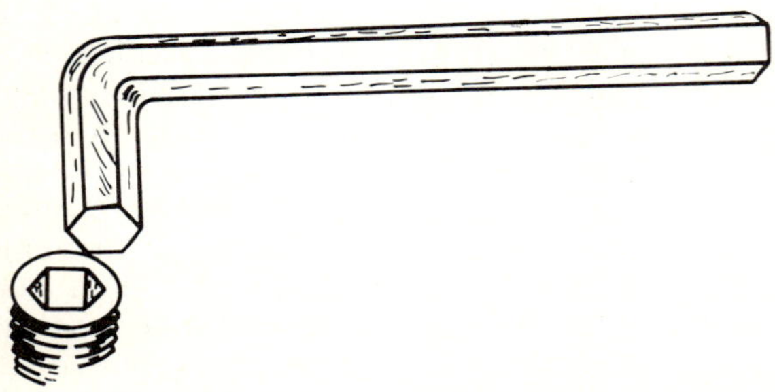

Fig. 13-5. Allen screw and wrench.

_____ **16** The proper screwdriver should fit the screw slot. The tip should be slightly narrower than the diameter of the screwhead and should fit the slot rather snugly.

_____ **17** Select a proper-size screwdriver (see Step 19 below), and turn screws into and out of tapped holes in blocks. When using a screwdriver, guide the tip into the screw slot with one hand, and turn the handle with the other hand. Press the tip firmly and squarely into the slot. Turn the handle with one hand while holding the shank of the screwdriver between the fingers and thumb of the other hand. When the screw turns easily, spin the screwdriver shank between the fingers and thumb to speed up the job.

_____ **18** A screwdriver should not be used as a pry bar or chisel, since this can ruin it.

_____ **19** Remember that the standard screwdriver tip has practically parallel sides. If the sides were tapered, the tip would tend to rise up out of the screw slot as it was turned.

_____ **20** The screwdriver should be held straight in the slot, on a line with the screw. Holding it at an angle causes the blade to jump out and damage the screw slot.

True-or-False Test — Job 13

1 The sides of a standard screwdriver blade tip should be practically parallel. **1** _____

2 The square shank on a heavy-duty screwdriver can be gripped with a wrench to apply extra turning effort. **2** _____

3 Phillips-head screws have curved slots to keep the screwdriver from slipping out. **3** _____

4 Offset screwdrivers have blade tips parallel to each other. **4** _____

5 When a screwdriver tip becomes worn, it should be reground. **5** _____

Written Checkup — Job 13

1 List the cautions to be observed in using a screwdriver.

2 How are standard screwdrivers classified?

3 What is the difference between a Phillips and a Reed and Prince screwdriver?

NAME DATE CLASS

Objective: To learn about the types of hammers and their uses.

Equipment: Scrap sheet metal.

Tools: Various hammers, workbench, vise.

Job 14

HAMMERS

Safety: *Use care in handling scrap sheet metal. Avoid cutting your hands on the sharp edges. Make sure the hammer head is firmly attached to the handle, before using it. Never strike a hardened steel surface with a steel-head hammer. Particles of steel may break loose from the hammer or the surface and fly off at high speed. Such particles could easily put out an eye or cut a finger.*

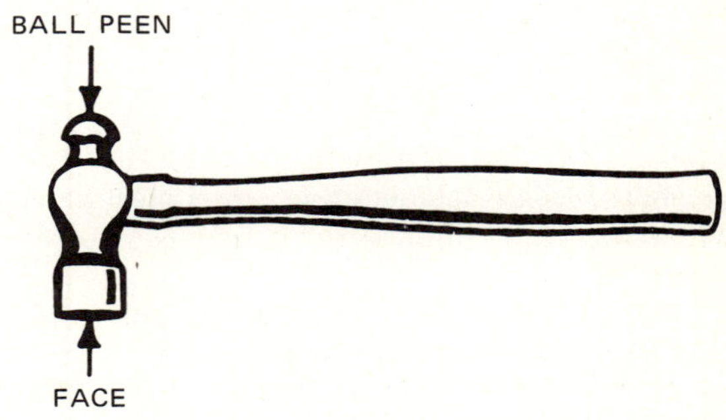

PROCEDURE

_____ 1 A variety of hammers are used in the shop (Fig. 14-1). You are probably familiar with the claw hammer, which has a claw, for pulling nails, opposite the driving head. This is a carpenter's tool; it is not of much use in machine shops. Instead, the ball-peen hammer is used (Fig. 14-2). When using this hammer, grip the handle firmly near its end. Swing the hammer so that the face strikes the object squarely (Fig. 14-2, lower illustration).

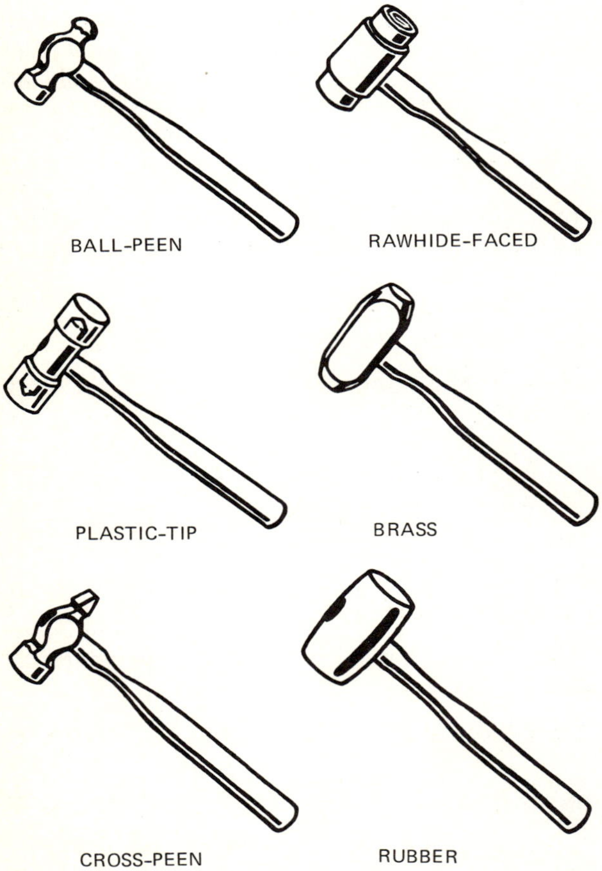

Fig. 14-1. Types of hammers.

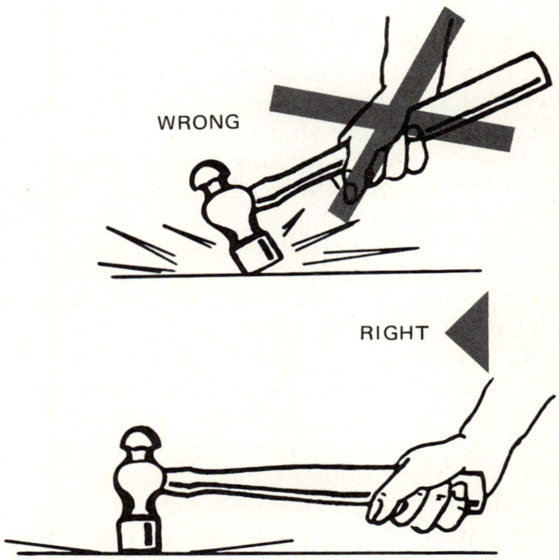

Fig. 14-2. Ball-peen hammer, showing wrong and right way to grip and use it. (*General Motors Corporation*)

_____ 2 If the work may be dented or otherwise damaged by the hard face of a ball-peen hammer, then a soft-face hammer should be used (Fig. 14-1).

_____ 3 Check the head of the hammer occasionally to make sure that it is firmly in place on the handle. A wedge or screw is used to spread the handle and tighten it in the eye of the hammer head

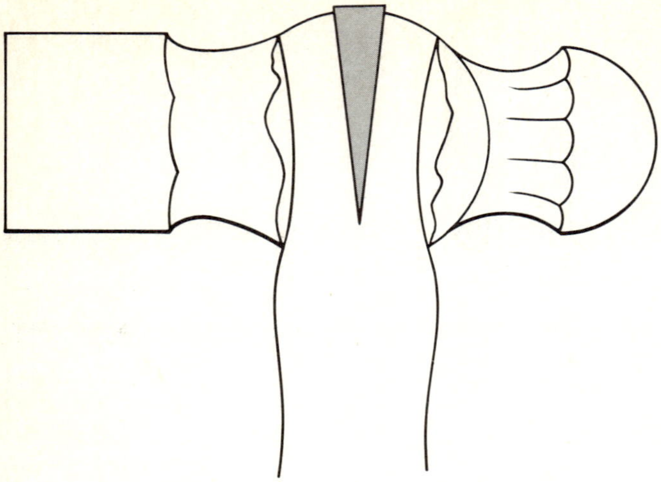

Fig. 14-3. Wedge installation in a ball-peen hammer.

(Fig. 14-3). If the wedge or screw has come loose, it should be driven tight. Someone could be very seriously hurt if the head flies off when the hammer is swung.

_____ **4** Plastic-tip hammers (Fig. 14-1) have soft plastic heads. On many hammers of this type, the heads screw in or are pressed in and can be replaced easily. A plastic-tip hammer is used when it is necessary to strike a part that might be damaged by a blow from a steel-head hammer. Damage can be prevented by striking the part with a hammer that is softer than the part itself.

_____ **5** Brass hammers (Fig. 14-1) are softer than plastic-tip hammers. They are very useful for working on aluminum and air-cooled engines. Brass hammers can be used on any material that may be damaged by striking with a steel-head hammer.

_____ **6** Rubber mallets (Fig. 14-1) are the softest type of hammer. A mallet is a short-handled tool; it, too, is used when a steel-face hammer would damage the piece being struck. Mallet heads are always made of a soft material, such as rawhide, rubber, or wood. Never use a mallet to drive nails or screws, as such objects will damage the mallet face. Rubber mallets are used in the automotive shop to drive on hub caps and wheel covers. The rubber mallet is soft enough not to damage the finish.

_____ **7** Never use a wooden hammer handle as a driver to assemble parts; and never use the hammer handle as a pry bar. The handle could split, and a split handle can cut or severely pinch your hand. Do not attempt to repair a split handle by binding it with string or wire. A split handle must be replaced.

_____ **8** Never strike a hard steel surface with a steel hammer. Small pieces of steel may break off and injure someone or damage the work. Punches and chisels may be struck with a hammer, because the heads of these tools are softer than a steel hammer head. However, punches and chisels must be kept properly dressed, to prevent small pieces from flying off a mushroomed head.

_____ **9** Hammers are classified according to weight without the handle. They are also classified according to style and shape (Fig. 14-1). In the automotive shop, the ball-peen hammer is the most used type. Ball-peen hammers are made in weights from 4 oz (ounces) to 2 lb (pounds) [113.4 g (grams) to 0.907 kg (kilogram)]. The 12-oz [340.2-g] hammer is often used by mechanics as a general all-purpose shop hammer.

_____ **10** Examine various hammers (Fig. 14-1). Note the relative weight and hardness of each, by tapping their heads lightly against the workbench surface.

_____ **11** Note the correct way to hold a hammer (Fig. 14-2). Grasp the handle firmly near the end, and swing from the elbow or the shoulder. For most work, the wrist is held almost rigid. Medium blows are produced by swinging from the elbow. Heavy blows result when the swing involves the whole arm.

_____ **12** Select a medium-weight ball-peen hammer and a crumpled scrap of sheet metal. Make sure that the hammer head is firmly attached to the handle (Fig. 14-3). Lay the sheet metal on the workbench surface, and try to smooth it with the hammer. Experiment with light, medium, and heavy blows. Note that the metal is badly dented if the hammer face does not strike the sheet-metal surface squarely.

_____ **13** Try to hit the sheet-metal surface squarely with the hammer face on every blow. Practice until you can do this. Then reverse the hammer, and note the effect of the ball peen on the metal. The ball peen is used for riveting and other such jobs.

_____ **14** Try rawhide, brass, wooden, plastic, and other hammers on the sheet-metal surface. Note how they act, and note their effects on the sheet metal.

_____ **15** When you have to hammer on a finished surface, protect the surface by holding a piece of wood, brass, lead, or other soft metal against it. Then hammer against the wood or soft metal, instead of directly on the finished surface.

True-or-False Test — Job 14

1 A hammer should be swung from the wrist for heavy blows. **1** _____

2 The hammer head is held in place on the handle by a wedge or screw. **2** _____

3 The hammer handle should be grasped near the end. **3** _____

4 The hammer handle makes a good pry bar and a mallet for bumping bearings into place. **4** _____

5 A mallet is a soft-face hammer. **5** _____

Written Checkup — Job 14

1 What is the purpose of the wedge or screw in the hammer handle?

NAME DATE CLASS

Objective: To learn about the types of pliers and their uses.
Equipment: Cotter pins, nuts, bolts, and wire.
Tools: Various pliers.

Job 15

PLIERS

PROCEDURE

_____ **1** Pliers (Fig. 15-1) are a special form of adjustable wrench. The jaws are adjusted as the two legs of the handle are moved on the pivot. Thus the jaws can be widened to grip an object, and to hold or turn it. Pliers should not be used on nuts or bolt heads. They will round off the edges of the hex and roughen the flats, so that a wrench will no longer fit onto the head properly.

_____ **2** Some pliers have a side cutter that can be used to cut wire (Fig. 15-1). There are also nippers, with cutting edges instead of pliers-type jaws. These can be used to cut wire, thin sheet metal, and small bolts.

_____ **3** There are many different sizes and shapes of pliers. In general, pliers are used for holding parts that you might not be able to hold in your hands. The most commonly used pliers are the combination pliers (Fig. 15-1). The slip joint in the jaw allows it to open wider to hold larger work. Most mechanics keep a 6-in [152.4-mm] combination pliers in their toolbox.

Combination pliers are a holding or gripping tool, not a turning tool. Do not use pliers to turn nuts. As you can see in Fig. 15-1, combination pliers have serrations or fine teeth cut in their jaws. These serrations can damage a nut or bolt by rounding off the corners in just a few seconds.

_____ **4** Long-nose pliers (Fig. 15-1) are usually called *needle-nose* pliers around the shop. They are very useful for gripping small objects, such as pins and clips. However, needle-nose pliers can be damaged by gripping too tightly. The long, thin jaws are easily sprung, broken, and nicked. Once needle-nose pliers are damaged, they become practically useless.

_____ **5** Diagonal pliers (Fig. 15-1) are commonly called *dikes* around the shop. They have a sharp cutting edge, which makes them useful for cutting wire and cotter pins. Diagonal pliers are a cutting tool. They should not be used to hold objects. Diagonal pliers are available in several sizes. The size is determined by the overall length of the pliers.

_____ **6** Channellock pliers (Fig. 15-2) are a holding tool similar to combination pliers. Channellock pliers have extra-long handles, which make them a very strong gripping tool. They are constructed differently from combination pliers. Channellock pliers have grooves on one jaw and lands on the other. To change the maximum opening of the jaws, the positions of the grooves and lands are changed. The groove-and-land design permits the jaws to be parallel to each other at any setting. This makes Channellock pliers less likely to slip when gripping an object.

_____ **7** Vise-Grip pliers are a well-known and widely used locking pliers (Fig. 15-3). This type of pliers can be locked onto an object. The screw at the end of the handle can be used to adjust the size of the jaw opening. The Vise-Grips lock in place when the jaws are closed around the work. They are re-

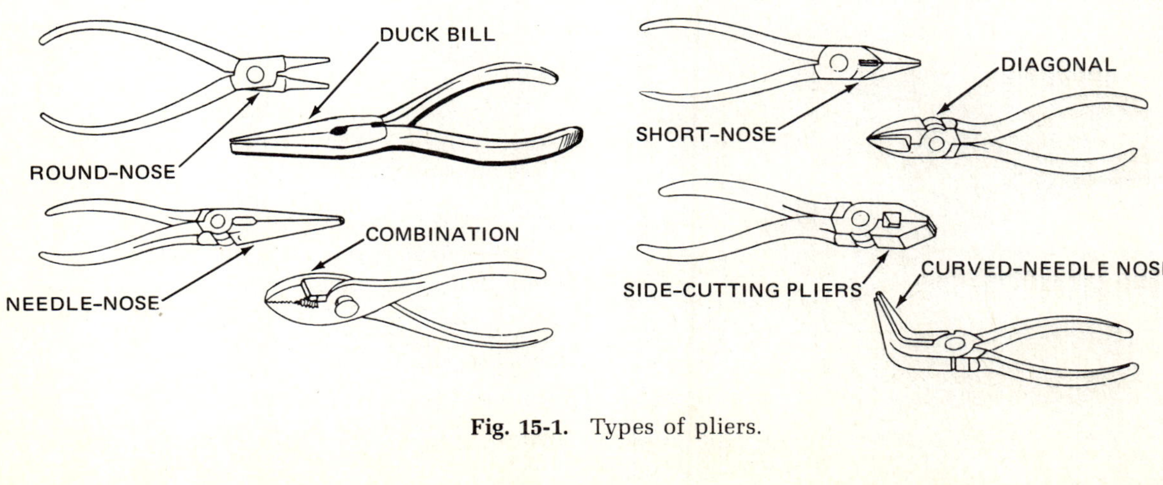

Fig. 15-1. Types of pliers.

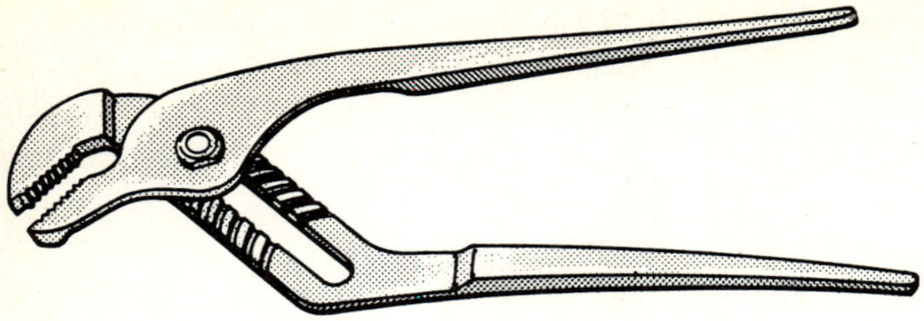

Fig. 15-2. Channellock pliers. (*Ford Motor Company*)

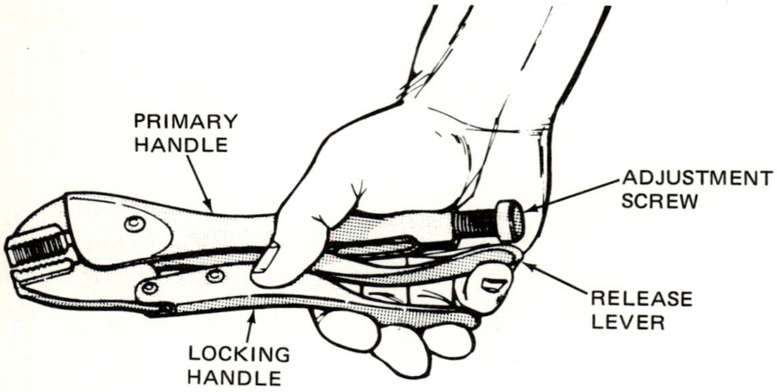

Fig. 15-3. Vise-Grip pliers. (*Ford Motor Company*)

leased by pulling on the release lever. Vise-Grip pliers can be used to grip nuts and bolts which have had their heads rounded off. However, they must not be used in place of wrenches.

_____ **8** Snap-ring pliers (Fig. 15-4) are of two types: inside pliers and outside pliers. The jaws on inside snap-ring pliers are like the jaws on combination pliers. That is, the jaws close and grip an object when the handles are closed. The jaws on outside snap-ring pliers open as the handles are brought together.

Snap-ring pliers must be used with care. Most snap rings are made of spring steel. If snap-ring pliers are used improperly, or not used at all to remove or install a snap ring, the snap ring can fly off and injure someone.

_____ **9** Examine various pliers (Fig. 15-1). Note how the different types are designed for gripping or for cutting. Note the various positions of the cutting edges (end, diagonal, and side). Also, note how the gripping jaws have serrations, or teeth, for a firm grip.

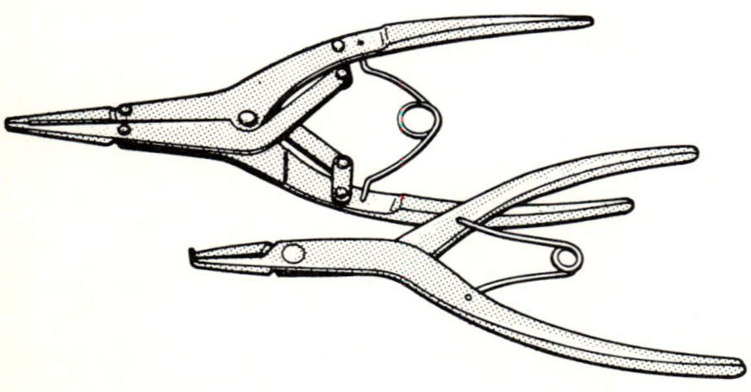

Fig. 15-4. Snap-ring pliers. (*Ford Motor Company*)

_____ **10** Use cutting pliers to cut some wire. Note the action of the cutting edges as they come together. Put a castellated nut on a drilled bolt. Install a cotter pin through the castellations of the nut and the hole in the bolt. Use diagonal cutters to pull the cotter pin firmly into place, and to cut the excess from the ends of the cotter pin. Then use the cutters to bend the two ends of the pin in opposite directions along the sides of the nut. Note how this prevents the nut from loosening.

_____ **11** Place a nut or a bolt head in the gripping jaws of a pliers. Note how the teeth bite into the nut or bolt surfaces. This actually damages the nut or bolt head, so that it may become impossible to use the correct wrench on it. Never use pliers on nuts or bolts. Besides damaging the nut or bolt, the pliers may slip and allow you to bang your knuckles.

_____ **12** Open the jaws of an adjustable combination pliers. Use the slip joint to increase the opening between the jaws. Note that this makes it possible to grip larger objects.

_____ **13** Cautions to observe in using pliers include the following:

a. Never use pliers on nuts or bolt heads.

b. Never use pliers on hardened steel surfaces, since this dulls the jaw teeth.

c. Never use pliers to turn pipe or tubing connections. The proper wrench should always be used, to avoid damage to the part.

d. Never try to cut steel wire, bolts, screws, or sheet metal with ordinary cutters. This might nick the cutting edges and ruin the cutters.

_____ **14** If the teeth of pliers become dull, they can be sharpened with a triangular file, such as described in Job 19.

True-or-False Test — Job 15

1 Pliers should not be used on nuts or bolt heads. 1 ______

2 Cutters should be used to loosen a nut if pliers do not loosen it. 2 ______

3 Pliers teeth can be sharpened if they become dull. 3 ______

4 Pliers should not be used on hardened steel objects. 4 ______

5 You can use a hammer to strike against the handle when cutting small bolts with side cutting pliers. 5 ______

Written Checkup — Job 15

1 List the cautions to be observed in using pliers.
2 Describe when the teeth on pliers should be sharpened.
3 Explain how to properly clean and maintain pliers.

NAME DATE CLASS

Objective: To learn about the types of wrenches and their uses.

Equipment: Nuts and bolts of various sizes, engine or machine with nuts and bolts that can be loosened and tightened.

Tools: Various wrenches.

Job 16

WRENCHES

PROCEDURE

_____ **1** The bolt heads and nuts commonly used in machines and engines are of the hexagonal type. That is, they have six flats around the outer surface (Fig. 16-1). These flats permit the use of a wrench to turn the nut or bolt.

_____ **2** A typical wrench of the open-end type is shown in Fig. 16-1. In most, the open end is at an angle of 15° with the handle, as shown. A few special-purpose wrenches have 90° angles.

_____ **3** To use an open-end wrench, place the open end over the flats of the bolt head or nut (Fig. 16-2). The proper size of wrench fits snugly on the bolt head or nut (Fig. 16-2). Then pull the wrench to turn the nut or bolt. The wrench handle provides considerable leverage, so the bolt or nut can be tightened or loosened easily.

_____ **4** Box wrenches (Fig. 16-3) serve the same purpose as open-end wrenches. However, they are more useful in many cases, because the box fits down over the nut or bolt head and will not slip off. The most widely used box wrench is the *12-point* wrench. This means the box has 12 notches in it, spaced 30° apart. It can turn a nut or bolt even where there is a swing of only 15°. On some box wrenches, the head is at an angle of 15° to the handle (Fig. 16-4). This provides additional clearance for the mechanic's hand.

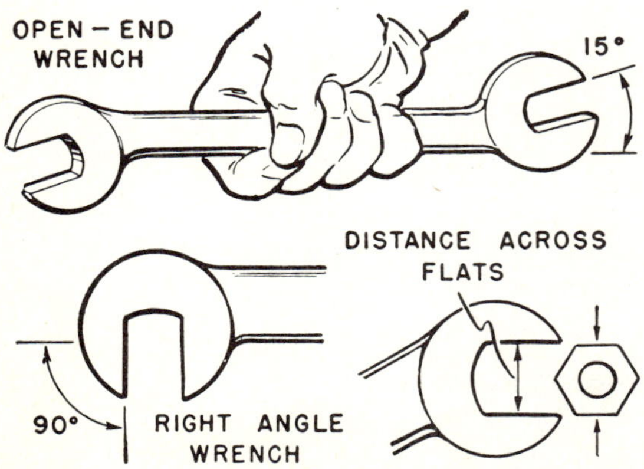

Fig. 16-1. Open-end wrenches.

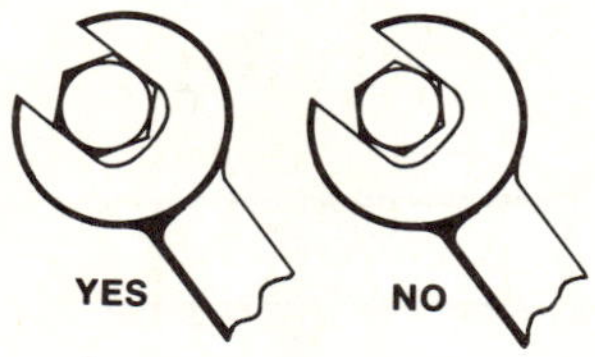

Fig. 16-2. An open-end wrench should fit the nut or bolt head securely.

_____ **5** One disadvantage of the box wrench is that it must be lifted off the nut and put back on in a different position with each swing of the wrench. This is a problem in a restricted work space. The open-end wrench slips onto the nut much more easily. Thus, many wrenches are made as combination open-end and box wrenches (Fig. 16-3). The box is used to break the nut or bolt loose. Then the tool is reversed, and the open end is used to back the nut off or the bolt out.

_____ **6** When using a wrench, it is usually better to pull on it than to push (Fig. 16-5). Sometimes, when the wrench is pushed, the nut or bolt gives way suddenly. You can bang your hand against the machinery before you catch yourself. This will not happen if you pull the wrench. Even when pulling, however, be ready for the nut or bolt to break loose suddenly. If it does this and you are not expecting it, you might pull the wrench off and hit yourself with it before you can stop.

_____ **7** The adjustable wrench (Fig. 16-6) has an adjustable jaw that can be moved back and forth to narrow or widen the distance between the jaws. Thus, the adjustable wrench can be made to fit many sizes of nut or bolt head. This tool is not intended to take the place of the standard wrench. However, it is handy for special jobs or odd sizes of nuts or bolts.

_____ **8** When using an adjustable wrench, make sure it is properly tightened on the nut or bolt head. Position the wrench so that the adjustable jaw is on the inside of the turning motion (Fig. 16-6). In this position, the pulling pressure keeps the adjustable jaw tight against the nut or bolt head.

_____ **9** Examine a set of open-end wrenches (Fig. 16-3). Note that wrenches with smaller openings have shorter handles. This reduces the chance of applying too great a force on a nut or bolt. Select several nuts, and find the proper wrenches to fit

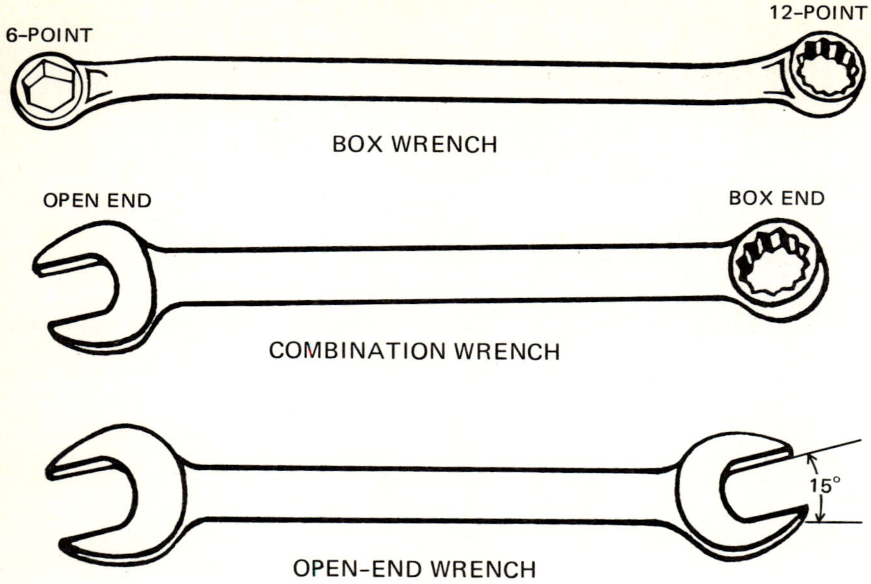

Fig. 16-3. Box, combination, and open-end wrenches.

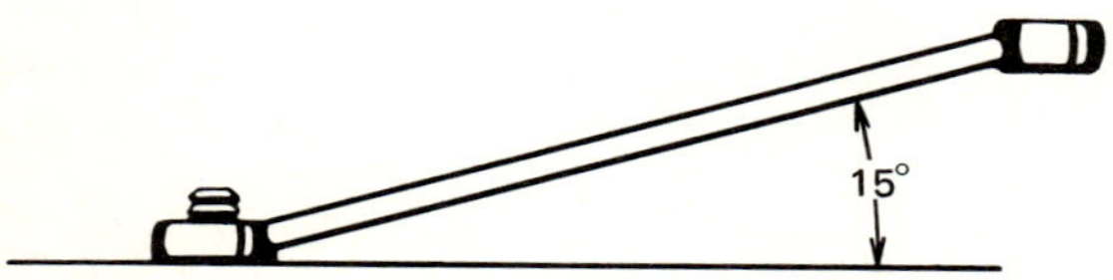

Fig. 16-4. Handle offset of a combination wrench.

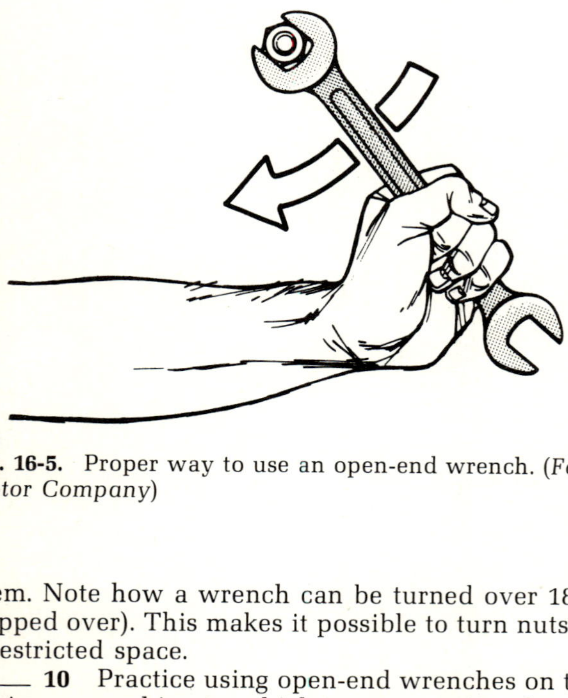

Fig. 16-5. Proper way to use an open-end wrench. (*Ford Motor Company*)

them. Note how a wrench can be turned over 180° (flipped over). This makes it possible to turn nuts in a restricted space.

_____ **10** Practice using open-end wrenches on the engine or machine to which you are assigned.

_____ **11** Observe the following cautions in using open-end wrenches:

a. Select the proper wrench to fit the nut or bolt head, to avoid a damaged wrench and nut or bolt.

b. When loosening or tightening nuts or bolts, do not strike the end of the wrench with a hammer or any other object. (The exception to this rule is the special striking wrench.)

c. Do not use a pipe or another wrench on the end of a wrench for additional leverage.

d. Pull, rather than push, on the wrench when loosening nuts or bolts.

_____ **12** Examine a set of box wrenches (Fig. 16-3). Select several nuts, and find the proper wrenches to fit them.

_____ **13** Practice using box wrenches on the engine or machine assigned to you.

_____ **14** The adjustable wrench has an adjustable jaw. This allows you to change the opening to fit the nut or the bolt head (Fig. 16-6). Examine an adjustable wrench. Turn the adjusting nut, and note how this moves the jaw. Select various bolts or nuts, and adjust the wrench to fit them.

NOTE: An adjustable wrench must be adjusted to the proper snugness on the nut or bolt before it is used. Otherwise, the wrench will be strained excessively and may be damaged. This strain may also damage the nut or bolt head.

_____ **15** Practice using the adjustable wrench on the engine or machine assigned to you. Note the correct and incorrect ways of applying pull to the wrench handle (Fig. 16-6). Maintain the pull so that most of the pressure is on the stationary jaw.

_____ **16** Keep adjustable wrenches clean, and the adjusting nut and jaw oiled.

True-or-False Test — Job 16

1 The length of an open-end wrench is not proportional to the size of its opening. **1** _____

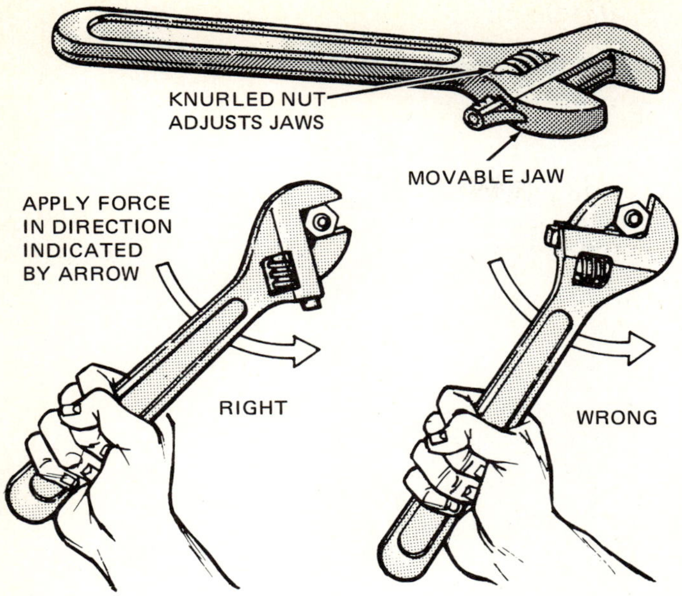

Fig. 16-6. When using an adjustable wrench, pull on the handle so the major load is carried by the stationary jaw, as shown in the lower left illustration. (*Ford Motor Company*)

2 A box wrench can be used with the ratchet and offset handle from a socket set. **2** ______

3 It is preferable, to pull, rather than push, on open-end wrenches. **3** ______

4 Since the jaws of the adjustable wrench are adjustable, it is all right to use a hammer to strike the handle when loosening a tight nut. **4** ______

5 A pipe should never be placed on the end of a wrench for additional leverage. **5** ______

Written Checkup — Job 16

List the various wrenches contained in the set of basic hand tools with which you are working. Opposite the list of wrenches, write down the major use for each wrench.

NAME DATE CLASS

Objective: To learn about the types of socket-wrench sets and their uses.

Equipment: Nuts and bolts of various sizes, engine or machine with nuts and bolts that can be loosened and tightened.

Tools: Various socket-wrench sets and accessories, such as a speed handle, ratchet handle, universal joint, sliding offset handle, and extensions.

Job 17

SOCKET WRENCHES

PROCEDURE

_____ 1 The socket set (Fig. 17-1) is one of the most versatile sets of tools that you can have in your toolbox. A socket wrench is made up by selecting a socket and attaching it to a driver, such as a ratchet or speed handle. As you can see in Fig. 17-1, the socket wrench is similar to the box wrench. However, the sockets are detachable from the handle, or driver. This allows a single handle to be used for many sizes of sockets.

_____ 2 Figure 17-1 shows a typical socket set. A complete socket-wrench set consists of several different types of drivers, extensions, a universal joint, and a variety of sockets. Many socket sets are packaged in a metal case (Fig. 17-1) with a special place for each piece in the set. By checking the case, you can see immediately which pieces are missing. In addition, the metal case makes the socket set easier to keep clean and to store without danger of losing pieces. With every tool in its place, you do not waste time trying to find a small tool in a large toolbox.

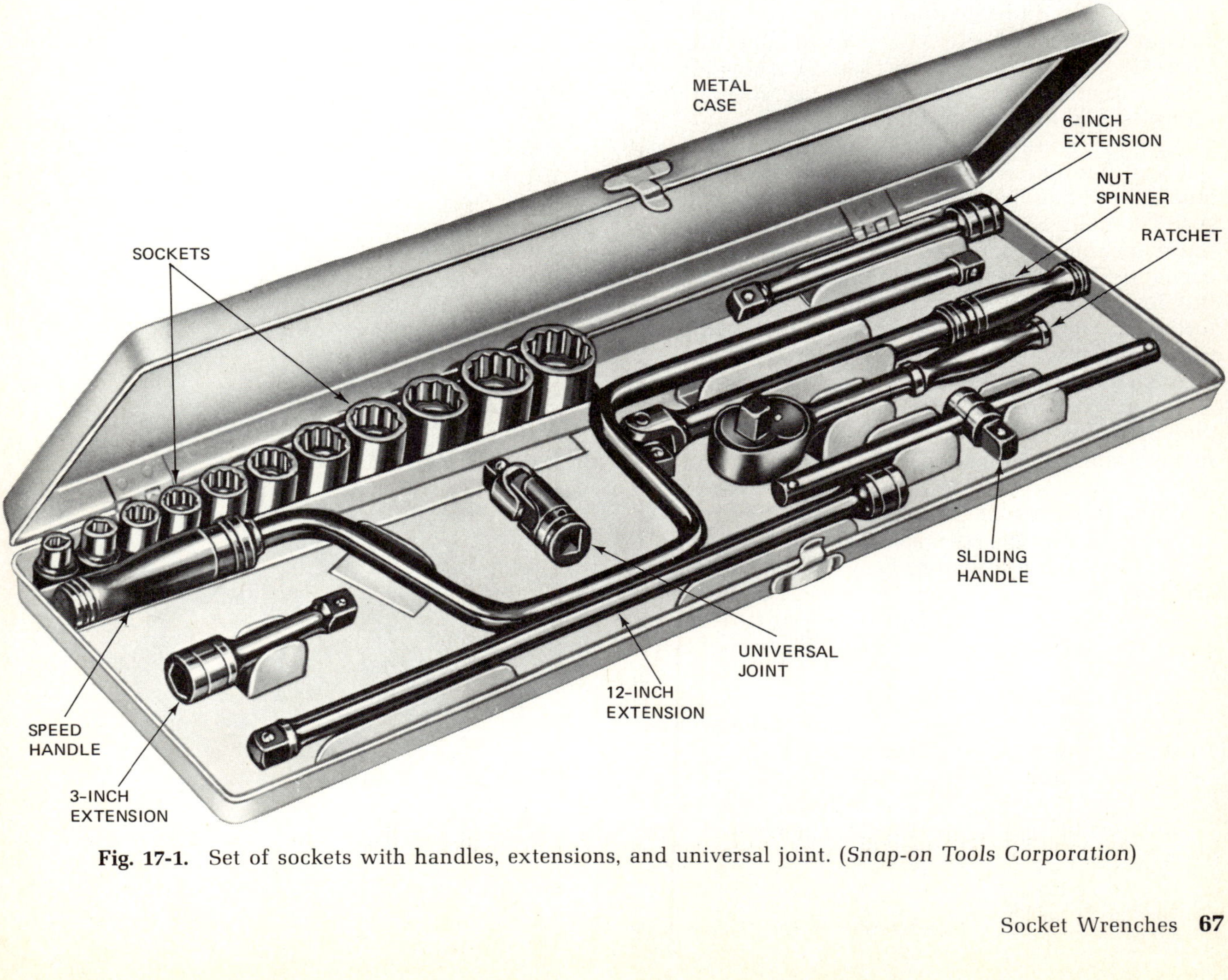

Fig. 17-1. Set of sockets with handles, extensions, and universal joint. (*Snap-on Tools Corporation*)

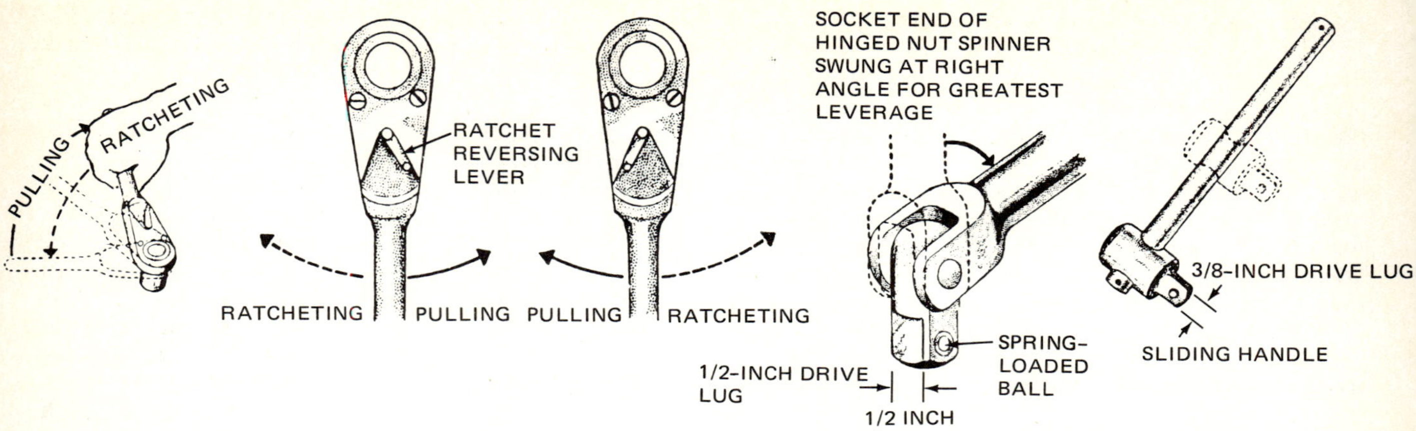

Fig. 17-2. Left, the ratchet handle will rachet in either direction, according to the position of the control lever. Right, use of the nut spinner and sliding handle.

_____ **3** Socket sets are widely used in automotive work. The size of a socket set is determined by the size of the square driving lug on the driver or handle (Fig. 17-2). Commonly used socket sets have $\frac{1}{4}$-in, $\frac{3}{8}$-in, and $\frac{1}{2}$-in drives. In a $\frac{3}{8}$-in set, for example, all the drivers have a $\frac{3}{8}$-in square driving lug, and all the sockets have a $\frac{3}{8}$-in square drive-end opening.

_____ **4** A variety of handles are used with sockets (Figs. 17-1 and 17-2). The ratchet handle can be swung back and forth without lifting the socket from the nut or bolt head (Fig. 17-2). The ratchet reversing lever (Fig. 17-2) can be switched for ratcheting in either direction. This allows you to use the handle to either loosen a nut or tighten it.

_____ **5** Figure 17-1 shows four frequently used types of handles. Besides the ratchet, these are the sliding or T handle, nut spinner or breaker bar, and speed handle. Each handle has special advantages; the mechanic soon learns which handle to use for each job. The speed handle is the fastest turning tool. But for breaking loose tight bolts and nuts, more torque can be applied with the breaker bar. In situations permitting only a small amount of turning through an angle of a few degrees, the ratchet is best.

_____ **6** The drive lug on each handle, as well as the drive-lug-end of an extension, contains a spring-loaded ball on one flat (Fig. 17-2). This ball fits into a recess, in the drive lug, which keeps it in position. When a socket is pushed onto the drive lug, the ball fits into a recess machined into the flat of the socket. This mating feature keeps the socket in the driver during normal usage. However, a slight pull on the socket causes the ball to be depressed back into the driver, so the socket can be disconnected from the driver.

_____ **7** The hinged handle or nut spinner (Figs. 17-1 and 17-2) is the handle that gives the greatest amount of leverage. To loosen tight nuts or bolts, swing the handle at right angles to the socket, as shown in Fig. 17-2. After the nut is loosened, the handle can be moved to the vertical position. The nut is then spun off with the fingers.

_____ **8** The sliding handle (Figs. 17-1 and 17-2) can be positioned anywhere along the sliding bar. Some jobs can be performed better with a T handle, especially when it is important not to lean the socket to one side. Many mechanics prefer to remove and install spark plugs with a T handle. The T handle keeps the socket from leaning against the spark plug and breaking it.

_____ **9** The speed handle (Figs. 17-1 and 17-3) works like a carpenter's brace and bit. After a nut or bolt is loosened with a ratchet, sliding handle, or breaker bar, the speed handle can be used to remove it quickly. However, the leverage of the speed handle is not great enough for breaking loose or tightening most nuts and bolts.

_____ **10** Figure 17-1 shows several other parts that increase the versatility of a complete socket set. Extensions of various lengths are used to increase the distance from the socket to the handles, when needed. Commonly used extension lengths are 3 in [76.2 mm], 6 in [152.4 mm], and 12 in [304.8 mm] (Fig. 17-1). Any number of extensions can be attached

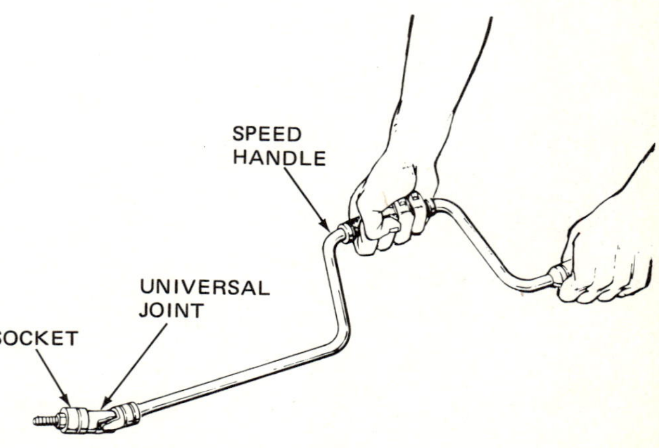

Fig. 17-3. Speed handle with universal joint and socket attached. (*Ford Motor Company*)

together to increase the distance from the socket to the handle, as required.

_____ **11** The universal joint (Figs. 17-1 and 17-3) is used when it is necessary to turn a nut while holding the handle at an angle. As you can see in Fig. 17-3, the universal joint is attached between the handle and the socket. For some jobs, the universal joint and socket may be too long and awkward to use. A one-piece, combined universal joint and socket is available.

_____ **12** Among the other available accessories are adapters that permit a handle of one size to be used to drive a socket of a different size. Allen wrenches, Phillips screwdrivers, and standard screwdriver blades of various sizes are available for use with socket-set handles.

_____ **13** Figures 17-1 and 17-4 show different sockets. A socket has a square opening in one end, to accept the drive lug on a handle. The other end of the socket is much like the opening in a box-end wrench (Job 16). Most sizes of sockets are available with either 6 or 12 points (Fig. 17-4). The 12-point socket needs to be swung only half as far as the 6-point socket before it must be lifted and refitted to the nut for a new grip. The 8-point socket (Fig. 17-4) is useful for turning square pipe plugs.

_____ **14** Sockets are available in different lengths, as well as with different sizes of opening. In the U.S. Customary System of measurement, nut and bolt-head sizes are usually graduated in sixteenths ($\frac{1}{16}$) of an inch. Therefore, sockets are also graduated in sixteenths of an inch.

Certain studs use nuts which cannot be reached with standard-length sockets. For these, and for spark plugs, special deep sockets are needed. Figure 17-4 shows the difference between a regular socket and a deep socket.

_____ **15** If you are going to work on imported cars and motor cycles, you will need a set of metric sockets. USCS-sized tools will not fit some nuts and bolts on imported vehicles. Study Fig. 17-5 to familiarize yourself with the difference between metric and USCS sizes of sockets.

_____ **16** When you work with metric nuts and bolts, remember that the bolt-head sizes are expressed in millimeters. But the drive lugs on socket drivers are measured in inches throughout the world. That is, the drive end of a socket is always made to accept $\frac{1}{4}$-in, $\frac{3}{8}$-in, and $\frac{1}{2}$-in drive lugs (Fig. 17-6).

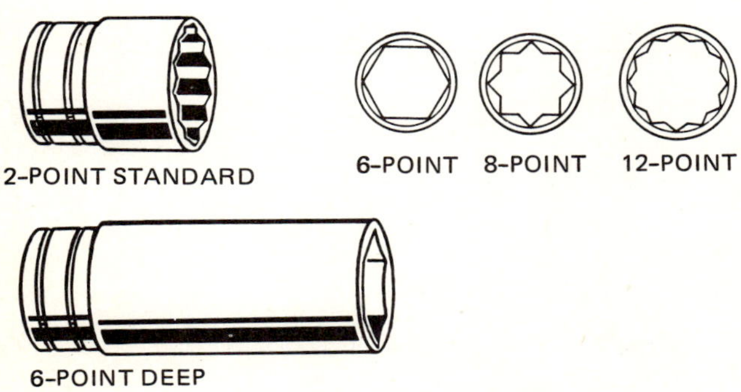

Fig. 17-4. Standard and deep sockets.

_____ **17** Examine a set of socket wrenches (Fig. 17-1). Note the various handles with which the sockets may be used.

_____ **18** Snap sockets onto the sliding handle, speed handle, and ratchet to get the feel of these handles (Fig. 17-2). Try as many handles as time and equipment permit. On the ratchet handle, move the ratchet reversing lever one way and then the other way. Note how the handle ratchets either way (Fig. 17-2).

_____ **19** Attach a universal joint to a driver such as a speed handle (Fig. 17-3). Then fit the socket to a bolt or nut to be turned. Note how the socket can be turned with the handle at an angle.

Multiple-Choice Test — Job 17

1 The fastest handle in a socket set is the **A** sliding handle **B** breaker bar **C** ratchet **D** speed handle **1** _____

1 A $\frac{1}{2}$-in drive socket can be made to fit on a $\frac{3}{8}$-in drive ratchet by use of **A** an adapter **B** an extension **C** a universal joint **D** a universal socket **2** _____

3 The size of a socket set is determined by the size of the **A** nut that the socket fits **B** length of the extensions **C** size of the driving lug on the handles **D** depth of the sockets **3** _____

4 Metric sockets use **A** $\frac{1}{4}$-in, $\frac{3}{8}$-in, and $\frac{1}{2}$-in drives **B** 6-mm, 10-mm, and 15-mm drivers **C** special metric drivers **D** special metric extensions **4** _____

5 Spark plugs require the use of **A** regular metric sockets **B** deep metric sockets **C** deep USCS sockets **D** standard USCS sockets **5** _____

Written Checkup — Job 17

1 Make an inventory of the socket set that you have in your toolbox. Write down the name and classification of each piece, for example, "$\frac{7}{16} \times \frac{1}{2}$ in 12-point regular socket."

2 Figure 17-5 shows USCS and metric socket sizes. List the sizes of sockets in your toolbox in a column. Then, next to each socket size, write the size of metric nut and bolt head that the socket will fit.

SOCKET	METRIC SIZE	DECIMAL EQUIVALENT IN INCHES		USCS SIZE	SOCKET	METRIC SIZE	DECIMAL EQUIVALENT IN INCHES		USCS SIZE
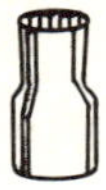	3 mm	0.118	0.125	1/8 INCH	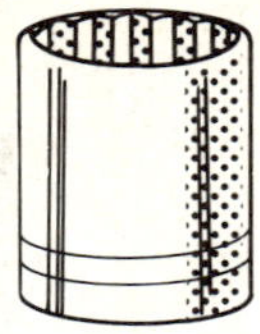	16 mm	0.630	0.625	5/8 INCH
	4 mm	0.157	0.187	3/16 INCH		18 mm	0.709	0.687	11/16 INCH
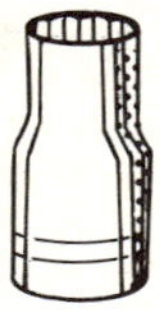	6 mm	0.236	0.250	1/4 INCH	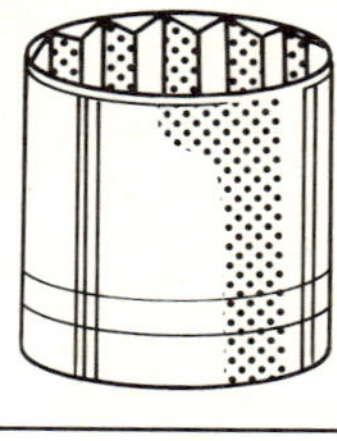	19 mm	0.748	0.750	3/4 INCH
	9 mm	0.354	0.312	5/16 INCH		20 mm	0.787	0.812	13/16 INCH
	10 mm	0.394	0.375	3/8 INCH	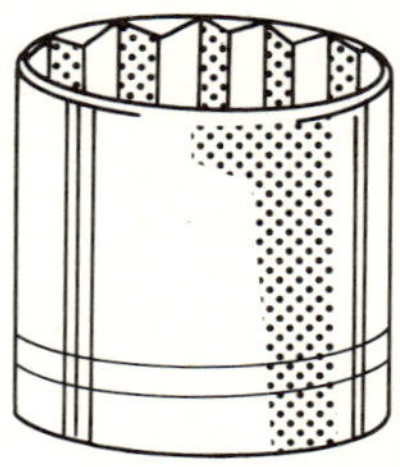	22 mm	0.866	0.875	7/8 INCH
	12 mm	0.472	0.437	7/16 INCH		24 mm	0.945	0.937	15/16 INCH
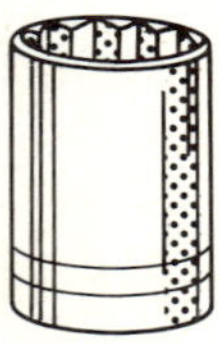	13 mm	0.512	0.500	1/2 INCH	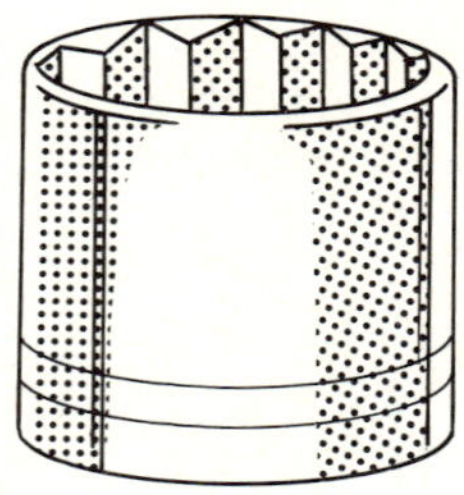	25 mm	0.984	1.00	1 INCH
	15 mm	0.590	0.562	9/16 INCH					

Fig. 17-5. Comparison of metric and USCS sizes of sockets. (*Dana Corporation*)

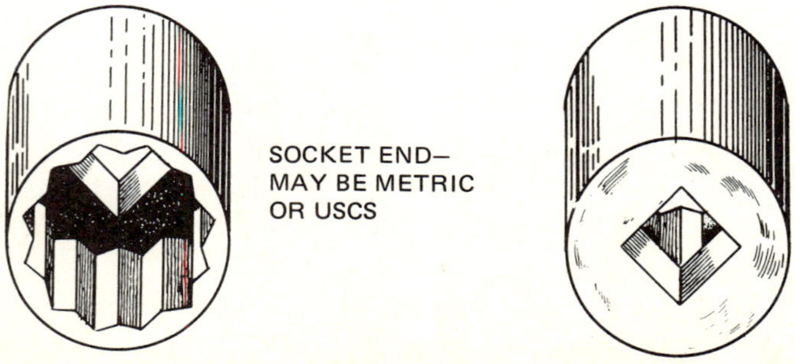

Fig. 17-6. The drive end of a socket is always made for the U.S. Customary System.

NAME DATE CLASS

Objective: To learn how to use and read torque wrenches.

Equipment: Engine or other assembly with different sizes of nuts and bolts.

Tools: Torque wrenches.

Job 18

TORQUE WRENCHES

PROCEDURE

_____ **1** The torque wrench (Fig. 18-1) is a special form of wrench. It has a measuring device that indicates the amount of torque or twist being applied to a nut or bolt. In much of today's machinery, engineers have carefully worked out the amount that bolts or nuts should be tightened. If they are not tightened enough, they may loosen in service. If they are tightened too much, the bolts or machine parts may be strained excessively and may break. In either case, the result could be severe damage to the machine. To prevent this, nuts and bolts must be tightened just the right amount. The torque wrench does this job.

_____ **2** On some torque wrenches, the amount of torque is registered on a dial or by a pointer as the handle is pulled (Fig. 18-1). On other types of torque wrenches, the specified torque is set by turning the handle (Fig. 18-2). A loud click is heard when that torque is reached. This type of torque wrench is called a *micrometer* or *clicker* torque wrench. Figure 18-2 shows how the specified torque is preset into the micrometer torque wrench by adjusting and locking the handle. This type of torque wrench permits very accurate tightening of nuts and bolts.

_____ **3** Regardless of their type, most torque wrenches measure torque in pound-feet (lb-ft). However, some torque wrenches (mainly those used for more accurate readings on smaller nuts and bolts) measure torque in pound-inches (lb-in). Twelve pound-inches equal one pound-foot. To convert pound-feet to pound-inches, multiply the pound-feet by 12. To convert pound-inches to pound-feet, divide the pound-inches by 12 (Fig. 18-3).

_____ **4** In the metric system, torque wrenches are scaled in kilogram-meters (kg-m) and in kilogram-centimeters (kg-cm). To convert pound-feet to kilogram-meters, multiply the pound-feet by 0.138. To convert pound-inches to kilogram-centimeters, multiply the pound-inches by 1.15. Several torque conversion formulas are shown in Fig. 18-3.

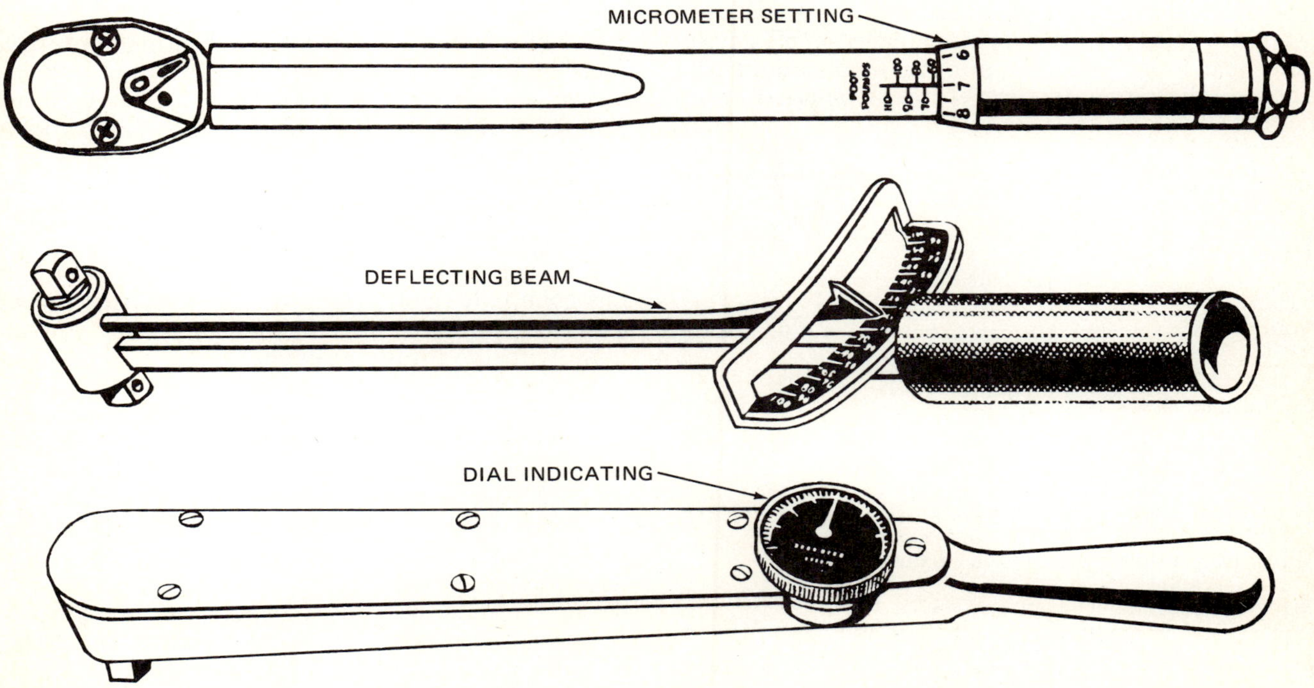

Fig. 18-1. Torque wrenches. (*General Motors Corporation*)

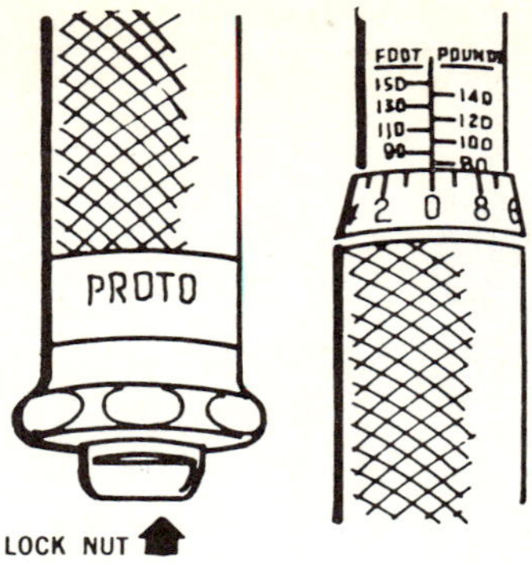

Fig. 18-2. Adjustment and locking of the micrometer type of torque wrench. (*Proto Tool Company*)

Fig. 18-4. Using a torque wrench to tighten cylinder-head bolts on a V-8 engine. The numbers indicate the order in which the bolts should be tightened. (*Snap-on Tools Corporation*)

_____ **5** To use a torque wrench, set the specification on the micrometer handle or note the specification on the scale or dial. Attach the proper socket to the torque wrench. Place the socket on the nut or bolt to be tightened (Fig. 18-4).

_____ **6** If an assembly containing many bolts or nuts is to be torqued, it is good practice to use a three-step procedure. First, torque all the bolts to about three-quarters of the specification. Recheck the specification and the torque-wrench setting. Then, apply the specified torque to each bolt. Last, retorque all bolts to the specification, to be sure that none were missed in the final torquing.

_____ **7** Use a smooth, steady pulling motion with a torque wrench. A fast or a jerky motion will cause the fastener to be tightened to an incorrect torque. Be sure to follow the manufacturer's torque specification for each fastener. In the specification charts, look for footnotes that indicate whether the threads are to be dry or lubricated with an antiseize compound. When you are tightening parts such as cylinder heads or intake manifolds, always find and follow the manufacturer's recommended bolt-tightening pattern. One manufacturer's cylinder-head bolt-tightening sequence is shown by the numbers in Fig. 18-4.

$$\text{Pound-inches} = \text{pound-feet} \times 12$$

$$\text{Pound-feet} = \frac{\text{pound-inches}}{12}$$

$$\text{Pound-feet} = \frac{\text{kilogram-centimeters}}{13.8}$$

$$\text{Pound-inches} = \frac{\text{kilogram-centimeters}}{1.15}$$

$$\text{Kilogram-meters} = \text{pound-feet} \times 0.138$$

$$\text{Pound-feet} = \text{kilogram-meters} \times 7.233$$

$$\text{Kilogram-centimeters} = \text{pound-inches} \times 1.15$$

$$\text{Kilogram-centimeters} = \text{pound-feet} \times 13.8$$

Fig. 18-3. Chart showing the formulas for converting torque specifications between the U.S. Customary System and the metric system. (*Proto Tool Company*)

_____ **8** Before selecting a torque wrench for a job, note the torque specification. Then select and use a torque wrench that will read about midrange for the specified torque. For example, a typical $\frac{1}{2}$-in-drive micrometer torque wrench (Fig. 18-1) measures from 30 to 200 lb-ft. It would be the best torque wrench for a job requiring about 100 lb-ft of torque. Typically, a $\frac{3}{8}$-in-drive torque wrench reads from 15 to 100 lb-ft. A $\frac{1}{4}$-in-drive torque wrench measures from 30 to 300 lb-in.

_____ **9** A torque wrench is a precision measuring instrument that accurately limits the torque applied to a fastener. Do not use the torque wrench as a general-purpose turning tool.

_____ **10** When using torque wrenches, observe the following:

a. Never move the handle of a micrometer torque wrench to a position below the minimum torque setting on the scale. However, when storing a micrometer torque wrench, always adjust the handle to its lowest setting.

b. Never use the torque wrench to tighten a fastener to a higher torque than the maximum reading on the scale.

c. Never use a torque wrench on a fastener that has been tightened with a wrench or socket. For an accurate torque reading, the final tightening must be done with the torque wrench.

d. A torque wrench has a recommended calibration interval. Most torque wrenches are guaranteed to be accurate within 4 percent. To ensure that the wrench is always this accurate, have it checked at the intervals specified by the manufacturer.

_____ **11** Examine and try torque wrenches (Fig. 18-4). Note how varying amounts of pull on the wrench handle register on the wrench dial. A torque wrench should be pulled evenly and steadily; it should not be jerked. Threads should be cleaned and lubricated before a nut or a bolt is tightened. This reduces thread friction and results in more accurate tightening.

CONVERTING MEASUREMENTS

1	______ lb-ft	______ lb-in	______ kg-m	______ kg-cm
2	______ lb-ft	______ lb-in	______ kg-m	______ kg-cm
3	______ lb-ft	______ lb-in	______ kg-m	______ kg-cm
4	______ lb-ft	______ lb-in	______ kg-m	______ kg-cm
5	______ lb-ft	______ lb-in	______ kg-m	______ kg-cm

______ **12** Take special care of the torque wrench, to avoid damaging it or ruining its adjustment. Keep it clean and oiled; put it away in a protected place after each use.

______ **13** Today, knowing how to convert measurements from the U.S. Customary System to the metric system is very important. Find a torque specification chart. In the left column above, write down five specifications given in pound-feet. Using Fig. 18-3, convert these specifications to pound-inches, kilogram-meters, and kilogram-centimeters.

Multiple-Choice Test — Job 18

1 A torque wrench is a **A** turning tool **B** precision measuring instrument **C** cutting tool **D** fastening device **1** ______

2 Another word for torque is **A** horsepower **B** weight **C** volume **D** twist **2** ______

3 When storing a micrometer torque wrench, you should adjust the handle to its **A** highest reading **B** most used position **C** off-scale position **D** lowest setting **3** ______

4 A good procedure to use with a torque wrench is to torque the bolts in an assembly in **A** one step **B** two steps **C** three steps **D** four steps **4** ______

5 To convert pound-feet to pound-inches, you multiply the pound-feet by **A** 12 **B** 0.138 **C** 7.233 **D** 1.15 **5** ______

Written Checkup — Job 18

Examine the torque wrenches available in your shop. Be sure you know how to set and read each type. On a sheet of paper for your notebook, list each type of torque wrench. Then write down how to check the calibration of each torque wrench and how to determine that a specified torque has been applied.

NAME DATE CLASS

Objective: To learn how to use punches.

Equipment: Metal blocks, both plain and with holes drilled through and pins driven tightly into holes.

Tools: Bench wire, punch set, hammer.

Job 19

PUNCHES

Safety: *Always wear safety glasses, goggles, or a face shield when working with a punch.*

PROCEDURE

_____ **1** A punch (Fig. 19-1) is a striking tool that is held in the hand and struck with a hammer. There are many different types and shapes of punches; each has its own special use. Punches are usually made of tool steel. They have a hardened point, but a soft striking end (to reduce the possibility of chipping from hammer blows).

_____ **2** Some punches are round in shape, with a knurled shank for better gripping with the hand. Other punches are square or octagonal in shape. These can be held with a gripping tool instead of in the hand. This prevents the punch from slipping and lessens the danger of injury to the hand.

_____ **3** Punches are used to knock out rivets or pins (Fig. 19-2). They are also used to align parts for assembly and to make the locations of holes to be drilled. The *drift,* or *starting, punch* is used with a hammer to loosen a pin or shaft (Fig. 19-2). The *pin punch,* which has a long thin shank, is then used to drive the pin out (Fig. 19-2).

_____ **4** The *center punch* is used for marking hole locations for drilling (Fig. 19-3). Unless the hole location is punched before a hole is drilled, the drill may wander (move around on the piece to be drilled). If the hole location is first center-punched, the drill will not do this (Fig. 19-4).

_____ **5** The center punch is also used to mark parts before disassembly, so they can be reassembled in the same relative positions. For example, it might be possible to put a cover plate on a housing in more than one position. However, before disassembly, a punch mark can be made on the edge of the cover plate, and another next to it on the housing. If the two punch marks are aligned on reassembly, the cover plate is reassembled in its proper position.

_____ **6** An aligning punch is usually about 1 ft [304.8 mm] long. It is very handy for lining up holes in mating parts during assembly.

_____ **7** Brass punches, or soft punches, are used to remove shafts, pins, and bearings. Usually, the brass punch is heavy enough to resist serious damage to itself, but soft enough not to damage a finished surface or chip a hardened surface.

_____ **8** Drift punches can be identified by their shape (Fig. 19-1). They have a long continuous taper from the point to the shank, or body. This is the punch to use first, to loosen a frozen bolt or pin. Once the bolt or pin is loose, a pin punch (Fig. 19-1) should be used to drive it out. As you can see in Fig.

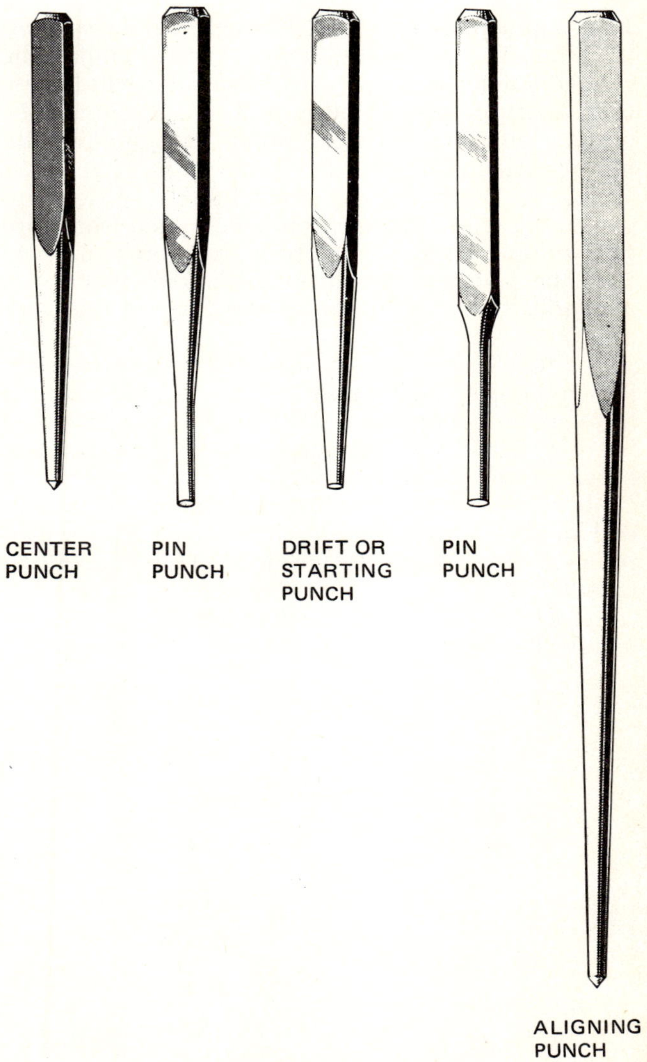

Fig. 19-1. Various kinds of punches. (*General Motors Corporation*)

Fig. 19-2. Using starting and pin punches. (*General Motors Corporation*)

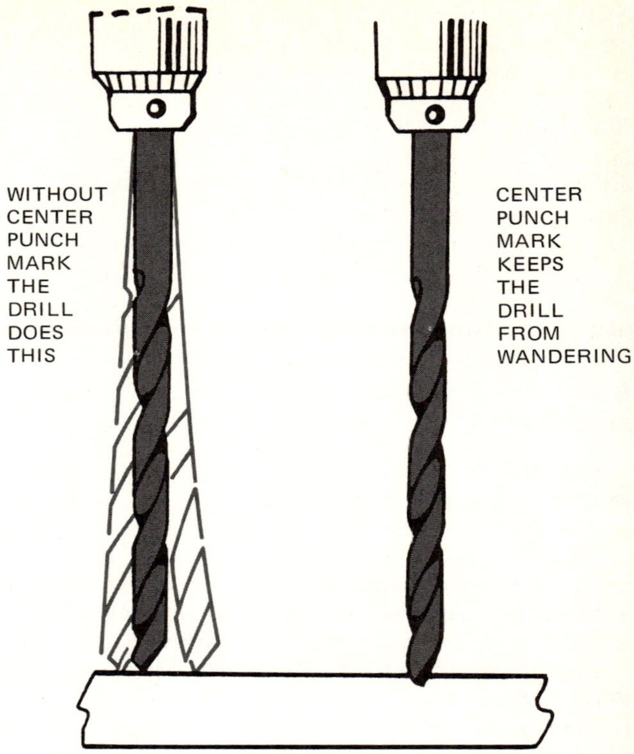

Fig. 19-4. How center-punching a hole location keeps the drill from wandering. (*General Motors Corporation*)

19-1, a pin punch can get through a hole without jamming. This is because the shank is ground to the same diameter as the tip for a considerable length. Both drift punches and pin punches should be ground with a circular, flat point. The point should not be edged or rounded.

_____ **9** When using a punch, hold it at right angles to the work. This prevents the punch from slipping sideways (Fig. 19-3). Strike the punch squarely with the hammer. Use a punch holder, when it is available and when there is room for it in the work area.

_____ **10** Examine a set of punches. Identify each punch. Before using a punch, check that it does not have a mushroomed head. When a mushroomed head is struck with a hammer, metal particles may break off and fly away at high speed. Such particles could cause serious eye injury or might be embedded in the skin. To avoid this danger, the mushroomed head should be ground off on a grinding wheel.

_____ **11** Place a small block of metal on the workbench. Take the center punch from the punch set.

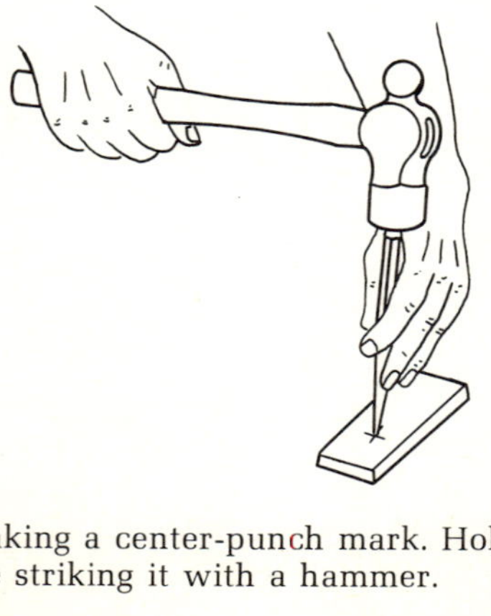

Fig. 19-3. Making a center-punch mark. Hold the punch vertical while striking it with a hammer.

Hold the center punch between your thumb and the ends of your fingers (Fig. 19-3), or hold it in a punch holder. Apply the point of the punch to a spot on the block. Make sure the punch is at right angles to the block surface.

_____ **12** Hold the hammer properly (Job 14). Bring the head of the hammer down, so it just barely touches the striking end of the punch. This helps you set up the hammer swing, so you will hit the end of the punch squarely. Now, lift the hammer and swing it down against the end of the punch (Fig. 19-3). Keep your eyes on the work, and not on the hammer head.

_____ **13** Lift the punch. Note the mark or depression made by the point of the punch in the metal block. Now, using a steel rule and a sharp-pointed scriber, scratch a line through the punch mark from one end of the block to the other. Then scribe several lines across the first line, at right angles to it. You now have several intersecting lines on the metal block.

_____ **14** Place the center punch on each intersection, and tap the punch with the hammer. Lift the punch, and examine each mark. If it is centered at the intersection of the lines, replace the punch point in the mark. Strike the punch sharply with the hammer to enlarge the mark. Repeat this procedure at each intersection, until you can center the punch mark perfectly at the point where the lines intersect.

_____ **15** Examine the point of the center punch periodically. After repeated hard usage, you may have to sharpen the point on a grinder (Job 30).

_____ **16** Put a metal block, with pins driven into it, in a vise. Select a drift punch (Fig. 19-1) slightly smaller in diameter than the pin you plan to remove from the block. Place the end of the drift punch on the end of the pin. Hammer on the punch to break the pin loose (Fig. 19-2). Do not use a center punch, since it will spread the end of the pin and make it harder to remove. (In actual practice, the pin would probably be peened over on the ends. Before it could be driven out, the peened-over part would have to be cut or filed off.)

_____ **17** Now select a pin punch (Fig. 19-1) slightly smaller in diameter than the pin. Place the punch end on the pin. Strike the punch with a hammer to drive the pin out.

_____ **18** Drift and pin punches should be ground and tempered when necessary. Always examine a punch carefully before using it.

_____ **19** Drift punches are usually made of brass or some other soft metal. They are essentially round rods. One end can be placed against a gear, shaft, or other part, and a hammer can be used on the other end. This directs the force of the blow exactly where it is wanted. It also protects the part from damage by direct hammer blows. Practice using a drift punch to get the feel of it.

True-or-False Test — Job 19

1 Before removing a pin, a center punch should be used to break it loose. **1** _____

2 It is dangerous to use a punch with a mushroomed head. **2** _____

3 When you use a punch, your eyes should be on the work, and not on the hammer. **3** _____

4 The pin punch should be slightly smaller in diameter than the pin that is to be driven out. **4** _____

5 The pin should be broken loose with a starting punch. **5** _____

6 Punches do not require tempering. **6** _____

7 Punches should be driven with a ball-peen end of the hammer. **7** _____

8 Drift punches are made of brass or other soft metals. **8** _____

9 You should watch the hammer head when driving on a punch, to make sure that it hits the punch end properly. **9** _____

10 If a center punch is used to break a pin loose, it will spread the end of the pin and make it harder to remove. **10** _____

Written Checkup — Job 19

1 Write down the procedure for using a center punch.

2 Write down the procedure for using punches to remove a pin.

NAME DATE CLASS

Objective: To learn about pullers, and how to assemble and use them.

Equipment: Various shafts and housings with gears, pulleys, bearings, and bushings to be removed.

Tools: Pullers and puller set.

Job 20

PULLERS

PROCEDURE

_____ 1 A puller set has many pieces that fit together in various ways to form different pullers (Fig. 20-1). A puller can be used to remove gears and hubs from shafts, bushings from blind holes, and cylinder liners from engine blocks. Pullers speed up many automotive service jobs. The proper puller will remove a gear from a shaft more quickly and with greater safety than a hammer and chisel or punch. In addition, the puller usually does not destroy the parts removed. They can be used again if they are otherwise okay.

_____ 2 There are three basic types of pullers: pressure screw pullers (Fig. 20-2), slide hammer pullers (Fig. 20-3), and combination pullers (Fig. 20-4).

_____ 3 The most common type of puller is the pressure screw puller (Fig. 20-2). It is so-called because the entire pulling force is produced by a pressure screw. When the puller is attached to the work, the pressure screw is placed against the shaft. As the screw is turned, it lengthens. This forces the cross arm or yoke of the puller away from the shaft. Thus,

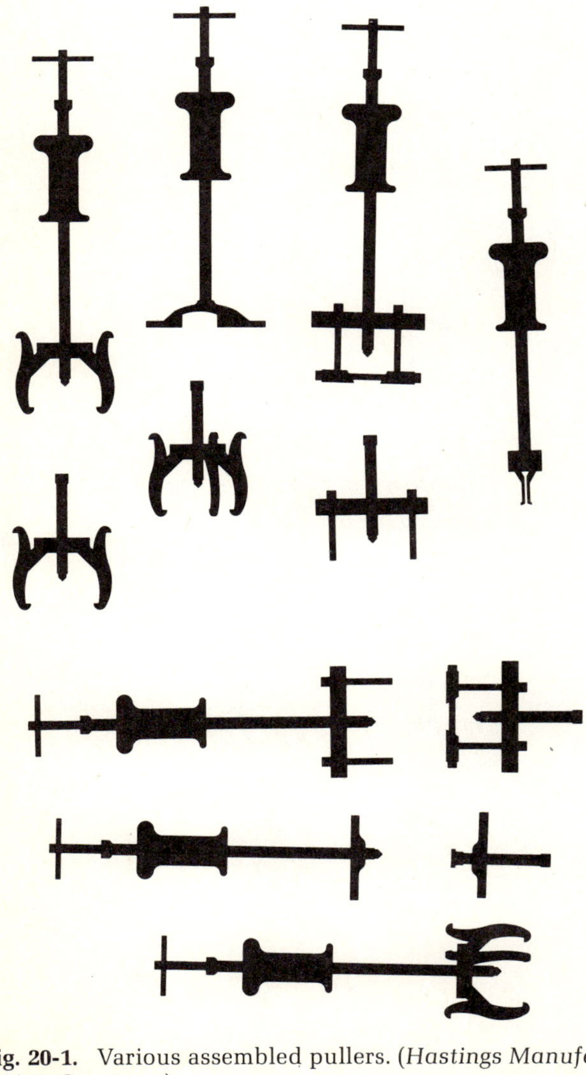

Fig. 20-1. Various assembled pullers. (*Hastings Manufacturing Company*)

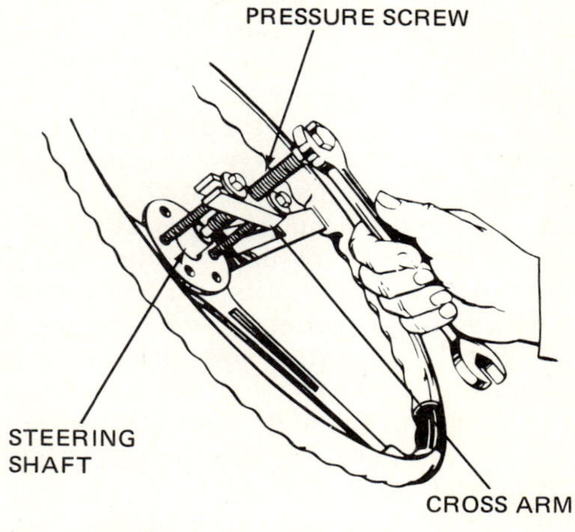

Fig. 20-2. Pressure screw puller being used to remove a steering wheel. (*Lisle Corporation*)

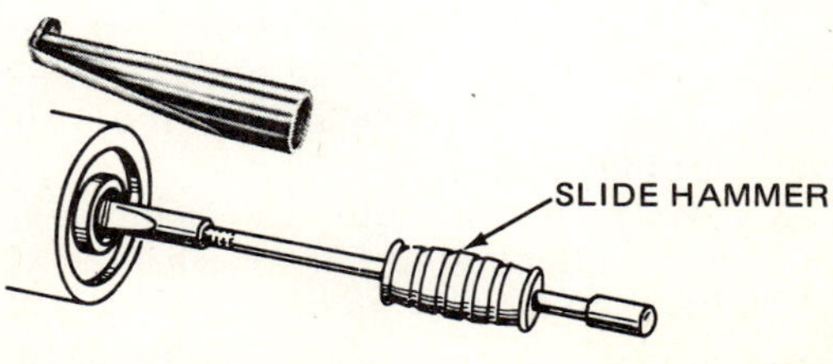

Fig. 20-3. Slide hammer puller being used to remove a grease retainer. (*Proto Tool Company*)

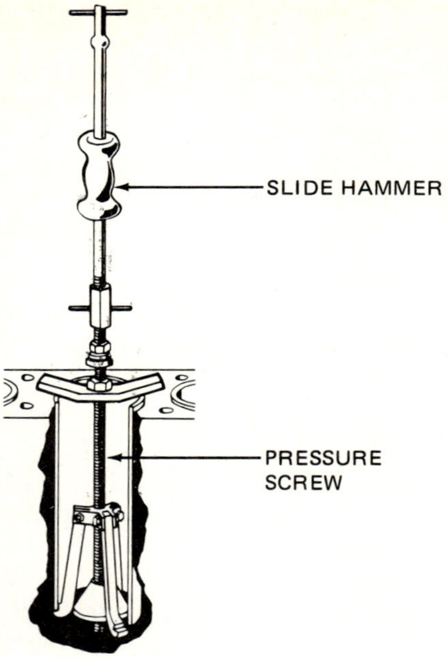

Fig. 20-4. Combination puller being used to remove a sleeve from an engine cylinder. (*Owatonna Tool Company*)

the pressure from the screw pulls the object from the shaft.

_____ **4** A slide hammer puller is shown in Fig. 20-3. This type of puller has a slide hammer that is used to strike the handle of the puller. As you can see in Fig. 20-3, the force of the blow is directed in a straight line against the object being pulled. This prevents the binding that can occur when an ordinary hammer is used to strike a puller. Slide hammer pullers are often used to remove hubs and axles, seals, and blind bushings (Fig. 20-3).

_____ **5** A combination puller is any puller that has both a pressure screw and a slide hammer. Figure 20-4 shows a combination puller set up to remove a cylinder liner from a diesel engine.

_____ **6** For large pulling jobs, hydraulic pullers are available (Fig. 20-5). A hydraulic puller uses a pressure screw. But instead of turning the screw to produce pulling power, a hydraulic pump is used to power a ram. The pressure screw is attached to the piston of the ram. As the piston moves, the screw pushes against the work. Tons of hydraulic pressure provide a hydraulic puller with a tremendous amount of pulling power.

_____ **7** To select a puller for any job, you must consider four factors. These are grip, push, size, and clearance. To pull an object, you must have a puller that can grip it. Figure 20-6 shows four different ways of gripping an object.

_____ **8** Jaw hooks or bolts are used to grip an object. They are attached to the cross arm or yoke of the puller. In *A* of Fig. 20-6, an *outside* grip is shown. The jaw hooks of the puller fasten around the outside of the object to be pulled. In *B*, jaw hooks are being used inside the object. This is called an *inside*

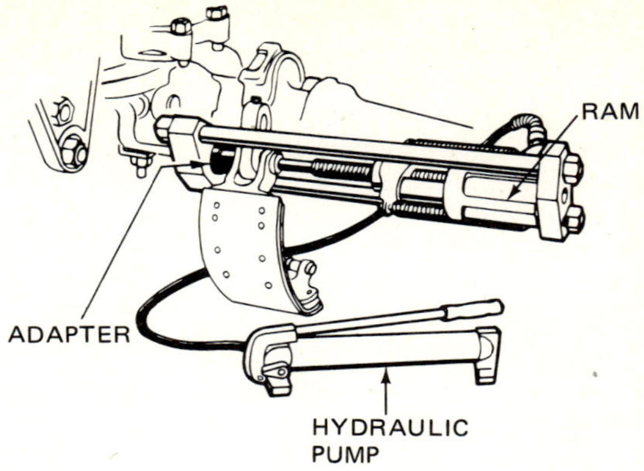

Fig. 20-5. Hydraulic puller being used to remove a brake anchor pin on a heavy-duty truck. (*Owatonna Tool Company*)

grip. Some pulleys and gears have tapped holes for puller attachment. In *C* of Fig. 20-6, you can see that bolts are placed through the puller cross arm and then threaded into the tapped holes. In *D* of Fig. 20-6, a bearing separator is being used as an adapter. The

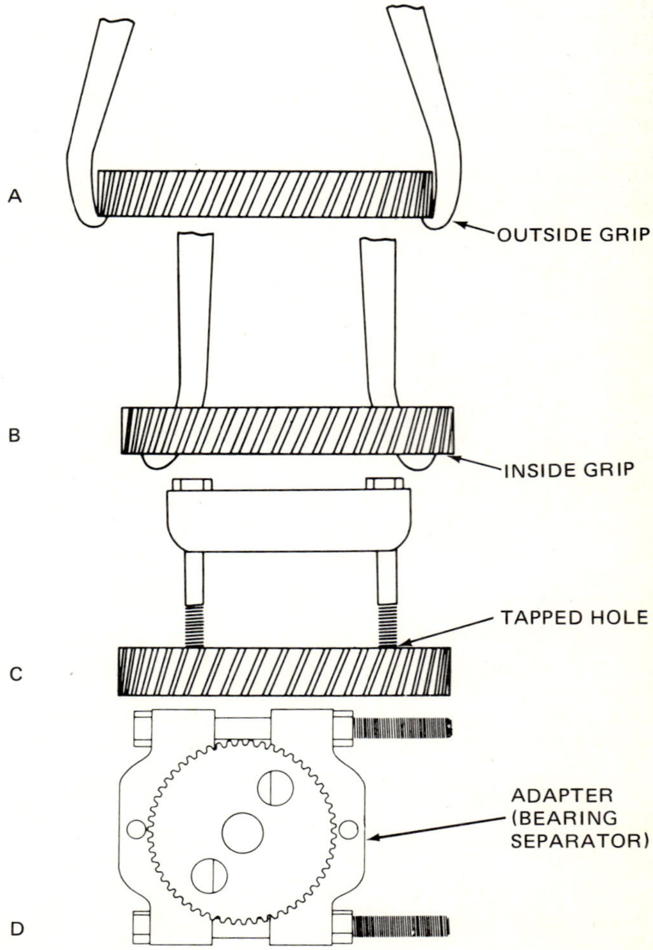

Fig. 20-6. Four types of grips that pullers use. (*Snap-on Tools Corporation*)

pull is made against the bearing separator. When the bearing separator is used this way, it acts as a backing plate. It distributes the pressure evenly across the back of the gear, thereby preventing damage to a soft or brittle gear.

_____ **9** To get pressure for a pull, you must push against something. Figure 20-7 shows different types of push. The most commonly used push is the center-point push. When a gear is removed from a shaft, the puller must push against the end of the shaft. This is shown in *A* of Fig. 20-7. Usually, the shaft has a slight countersink on its center line. A detachable tapered point is installed in the end of the puller screw and placed in the countersink in the shaft. This prevents damaging the shaft or the puller screw. It also aligns the puller screw with the shaft during the pull.

_____ **10** A *straddle yoke* must be used to remove a bushing or bearing from a housing that is open only on one side. This is necessary because there is nothing in the center to push against. The setup is shown in *B* of Fig. 20-7. A thin screw attachment is placed through the center of the bushing or bearing. A lip on the end of the attachment is worked behind the bushing or bearing. Then, as the handle on the screw is turned, the yoke comes to rest against, or straddling, the housing. Further turning of the handle pulls the bushing or bearing from the housing.

_____ **11** Sometimes an adapter (Fig. 20-7) can be placed in a hollow shaft to push against. When an adapter is used, the setup becomes a center-point push. Many puller sets have a variety of adapters which can be used when gears and pulleys must be removed from hollow shafts.

_____ **12** Occasionally you may have to remove a gear or bearing from a shaft or housing that has no center point to push against, while the opening cannot be straddled. For this type of job, the slide hammer puller can be used (Fig. 20-3).

_____ **13** Other factors must be considered when making up a puller from a complete puller set (Fig. 20-1). Most pulling jobs require the use of a puller with jaws set for an outside or an inside grip. When selecting jaws, you must be sure that they have adequate spread for the job, are wide enough, and have sufficient reach (Fig. 20-8).

_____ **14** Jaw spread is important because the jaws of the puller must spread wide enough to reach around the object to be pulled (Fig. 20-8). Unless the jaw spread is wide enough, the jaw hooks may not seat properly around the object. Then, during the pull the jaws may damage the object because of improper contact and slip off.

_____ **15** Jaws for pullers are available in several widths (Fig. 20-8). When pulling a gear, for example, you should not apply the pulling pressure against only a few teeth on the gear. This could break the teeth. For this type of pull you should use a wide jaw to spread the pulling pressure over as many gear teeth as possible.

_____ **16** Jaws with a long reach must be used to pull a gear that is set back from the end of a shaft (Fig. 20-8). When you set up a puller, check the distance from the end of the shaft to the back of the gear. Then be sure that the jaws on the puller have that much length, or reach.

_____ **17** When you set up a puller, check that the threaded length of the pressure screw is sufficient to

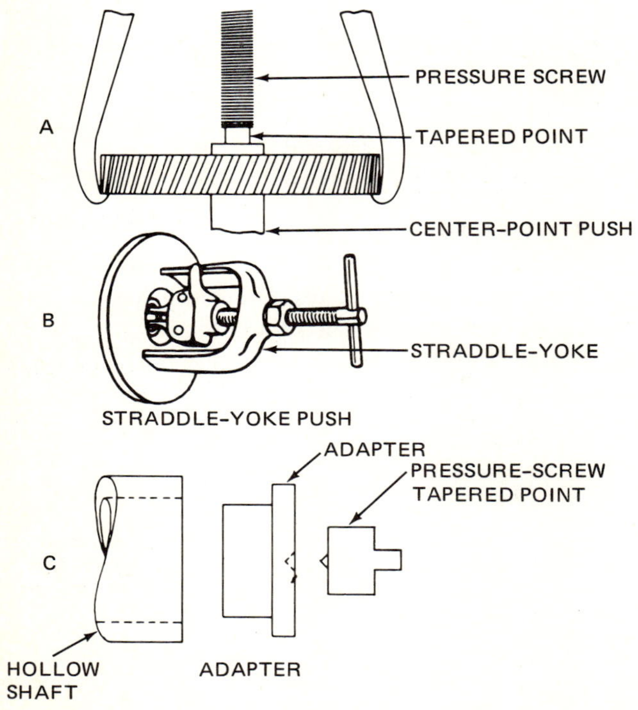

Fig. 20-7. Four methods of push that pullers can use. (*Snap-on Tools Corporation*)

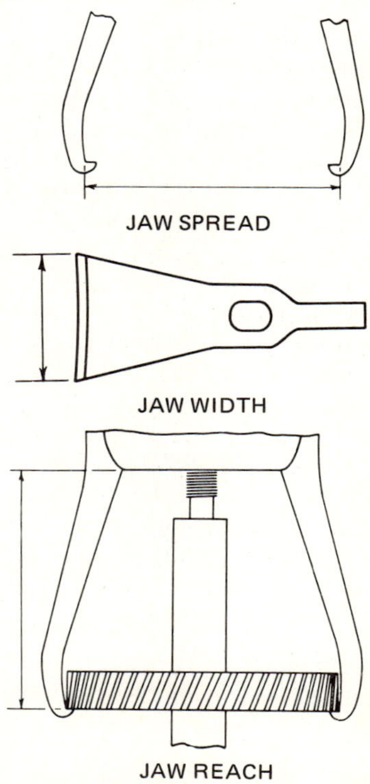

Fig. 20-8. Puller jaw dimensions. (*Snap-on Tools Corporation*)

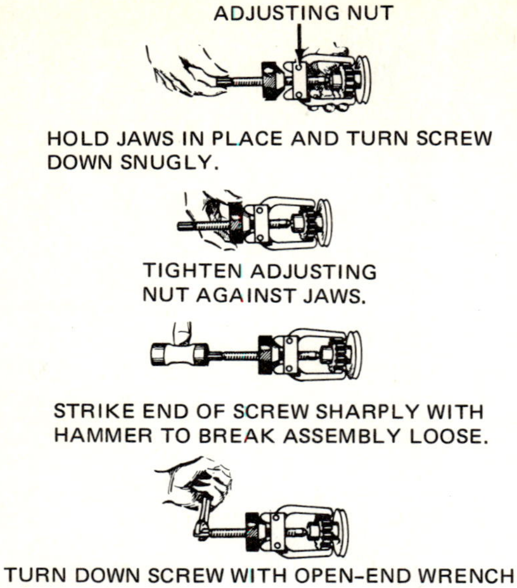

Fig. 20-9. How to use a puller. (*Proto Tool Company*)

complete the pulling job. If the pressure screw must be placed inside a hollow shaft, be sure the pressure-screw diameter is smaller than the diameter of the hole. A pressure screw that is too wide may become damaged and may even jam in the hole in the shaft. However, the pressure screw must be wide enough to handle the pressure that the pull will require. Always use the strongest puller that will fit the job. Faster and easier pulls can be made by keeping the threads on the pressure screw clean and coated with oil.

_____ **18** There must be sufficient clearance for jaw hooks to get around and behind the object being pulled. In addition, there must be sufficient room for the complete puller assembly, and you must have room to work. Pullers can create very high pressure. You must always be sure that the puller can be properly installed and used without interference from nearby parts.

_____ **19** Examine a puller set. Note the different lengths and sizes of jaws. Practice assembling different types of pullers. Put together a pressure screw puller (Fig. 20-2) and a slide hammer puller.

_____ **20** Place a shaft, with a gear or pulley to be removed, in a vise or on your workbench. Then assemble a puller to fit the job.

_____ **21** After the puller is selected or assembled, hold the jaws in place with your hand. Turn the pressure screw down snugly (Fig. 20-9).

_____ **22** Note, in Fig. 20-9, that many pullers use a tapered adjusting nut to hold the jaws in place. This prevents the jaws from slipping off during the pull. With your hand, tighten the adjusting nut against the jaws.

_____ **23** Strike the end of the pressure screw sharply with a hammer. This hammer blow causes the gear or pulley to break loose from the shaft.

_____ **24** Use a wrench to turn the pressure screw until the gear or pulley is free of the shaft.

_____ **25** Practice making pullers and pulling objects. If possible, make a puller using each of the four types of grip. Make a pull with each of them.

Multiple-Choice Test — Job 20

1 Three types of pullers are **A** hubs, axles, and seals **B** grip, push, and size **C** spread, width, and length **D** pressure screw, slide hammer, and combination **1** _____

2 The most common type of puller is the **A** slide hammer **B** combination **C** pressure screw **D** hydraulic **2** _____

3 To remove a seal from inside a housing, you would probably use a **A** slide hammer puller **B** pressure screw puller **C** combination puller **D** bearing separator 3 _____

4 When selecting a puller, you must consider push, size, clearance, and **A** type of puller **B** grip **C** jaw width **D** adapters **4** _____

5 For faster and easier pulls, the threads on the pressure screw should be **A** clean and oiled **B** ground off **C** covered with tape **D** welded closed **5** _____

Written Checkup — Job 20

Locate and familiarize yourself with all the pullers in your shop. Study the puller set. Be sure you can rapidly assemble each of the pullers shown. A great amount of time can be lost if you are unable to select and assemble the right puller quickly. On a sheet of paper for your notebook, make an inventory of the different pullers in your shop. List each type of puller and the lengths and sizes of jaws that are available for it.

NAME DATE CLASS

Objective: To learn about the types of chisels and their uses.

Equipment: Various pieces of flat and round soft steel stock, sheet metal, and blocks with peened-over pins or rivets.

Tools: Vise, chisels, hammer, steel rule, and scriber.

Supplies: Goggles, chipping shield.

Job 21

CHISELS

Safety: *When chipping, take care to protect your own eyes and the eyes of others in the room with you. You should always wear goggles, and others in the room should also wear them. If others are engaged in different activities, then you should set up a chipping shield. Flying chips will hit the shield and will not fly out into the room. These precautions will protect your eyes and the eyes of your fellow students. Always make sure that the hammer head is tight on the handle. Never use a dull chisel or a chisel with a mushroomed head. Such chisels should be ground and tempered.*

PROCEDURE

Part One GENERAL

_____ **1** The chisel has a single cutting edge and is driven with a hammer to cut metal. Several different shapes of chisels are shown in Fig. 21-1. Each shape has its special purpose, but the chisel most commonly used is the plain flat chisel.

_____ **2** The chisel should be held loosely in the left hand (Fig. 21-2), with the right hand swinging the hammer (or the reverse if you are left-handed). The reason for holding the chisel loosely is one of safety. If the hammer does not strike squarely or if it misses, the left hand will give with the blow and will be less subject to injury.

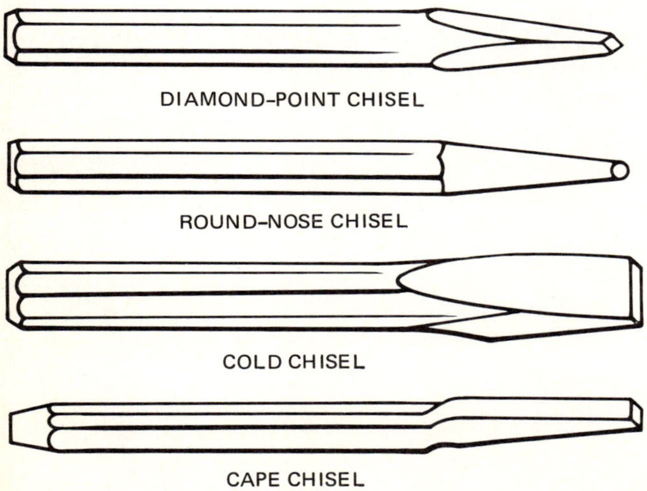

Fig. 21-1. Various types of chisels.

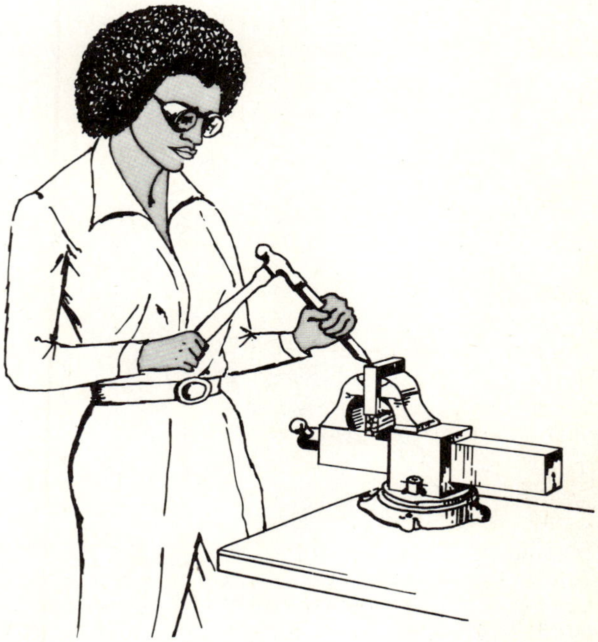

Fig. 21-2. Using a chisel and hammer.

_____ **3** A chisel that has mushroomed on the end (Fig. 21-3) because of repeated hammer blows should be dressed on a grinding wheel (Fig. 21-4). All the turned-over metal should be removed. Otherwise, this metal could break off and fly away as the hammer strikes it. It is difficult to be accurate with a

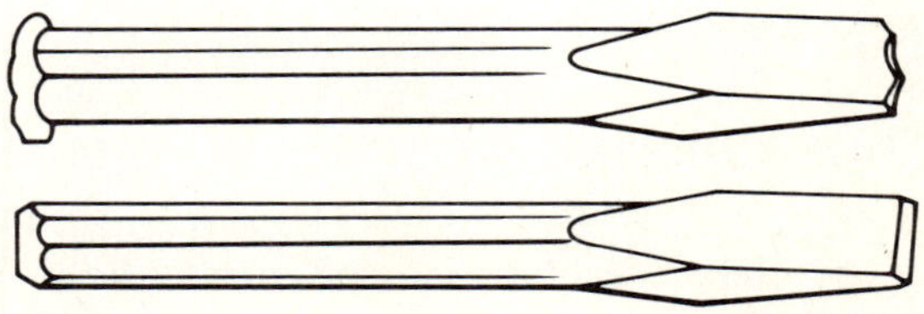

Fig. 21-3. Top, worn chisel with mushroomed head and dented cutting edge. Bottom, new or properly dressed chisel.

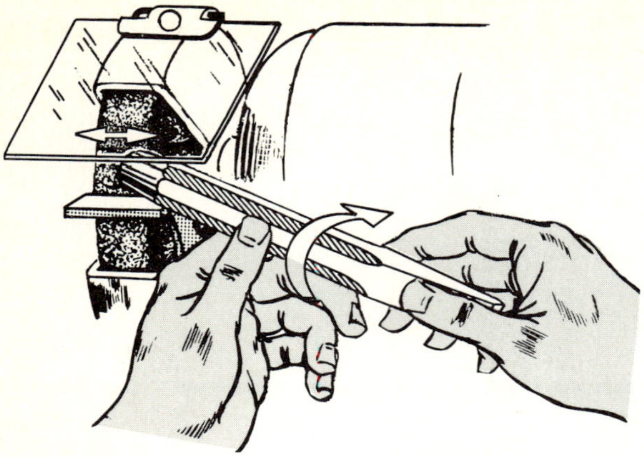

Fig. 21-4. Grinding the mushroom from the chisel head.

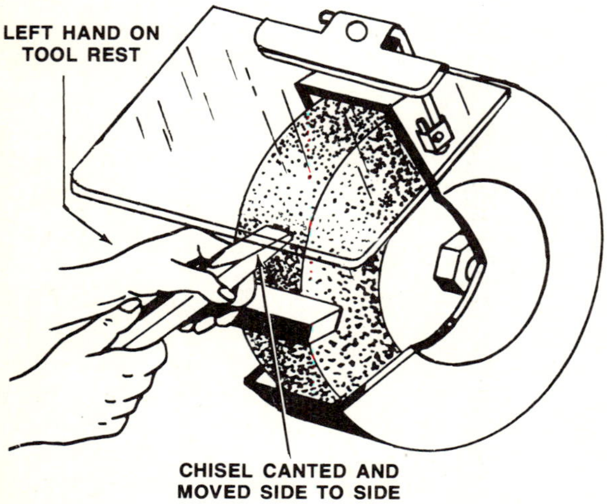

Fig. 21-5. Dressing or grinding a chisel. (*General Motors Corporation*)

mushroomed chisel and the flying chips might cut your hand or hurt someone else. Notice also how the cutting edge of the chisel should be dressed on the grinding wheel (Fig. 21-5).

_____ **4** Examine various chisels (Fig. 21-1). The flat chisel is used for cutting sheet metal and for chipping (removing metal from) flat surfaces. The cape chisel is used for cutting grooves or slots, such as keyways, and for chipping narrow, flat surfaces. The diamond chisel is used for cutting V-shaped grooves. The round-nose chisel is used for cutting round grooves.

Part Two CHIPPING WITH A FLAT CHISEL

_____ **1** Select a block of soft steel about $\frac{3}{4}$ in [19 mm] thick. Use a steel rule and scriber to mark a line all around the block, $\frac{1}{4}$ in [6.35 mm] from the edge (Fig. 21-6).

_____ **2** Put the steel block in a vise, with scribed marks above the vise jaws. Support the block with a piece of wood placed on the slide bar.

_____ **3** Put your goggles on, and properly place a chipping shield (if necessary). Select a flat chisel and a hammer. Begin chipping off the metal above the scribed line. Your job is to chip the block down to the scribed line.

_____ **4** To start the job, hold the chisel in one hand and the hammer in the other (Fig. 21-2). Place the edge of the chisel at one corner of the block, about $\frac{1}{16}$ in [1.6 mm)] below the top of the block. This will give you a $\frac{1}{16}$-in [1.6-mm] chip. Hold the chisel so that the lower face of the cutting edge is parallel to the top surface of the block. Note, in Fig. 21-7, that the chisel will cut upward and out of the work if it is held too low. It will cut down into the work if it is held too high. Note also (Fig. 21-7) what happens with a dull or improperly shaped chisel.

_____ **5** With the chisel held correctly, swing the hammer and start the chipping job. Keep your eyes on the cutting edge of the chisel, not on the hammer or the chisel end. After the cut is started, raise or lower the chisel to keep a uniformly thick chip coming off the work.

_____ **6** It is not necessary to keep the chisel pressed tightly against the chip all the time you are chipping. Pull the chisel back slightly every three or four hammer blows, and then press it up against the chip again just before the next hammer blow falls. This will rest your hand. Do not hold the chisel too tightly, or your hand will soon become tired. Also, keep your hand well away from the end of the chisel.

_____ **7** After the first cut, examine the surface; note any irregularities and chisel marks. These are mostly caused by failure to hold the chisel at a uniformly correct angle during the chipping.

_____ **8** Repeat Steps 4 to 7, trying to end up with a smooth surface. Then repeat the operation again, this time leaving only about $\frac{1}{32}$ in [0.8 mm] of material above the scribed lines.

_____ **9** Make the final cut carefully, at the scribed lines. Try to leave a smooth and finished surface.

_____ **10** Remember the following when chipping:

a. Always work with a sharp chisel. A dull chisel will not do good work.

b. Make the hammer blows uniform.

c. Keep the chisel at the proper angle constantly.

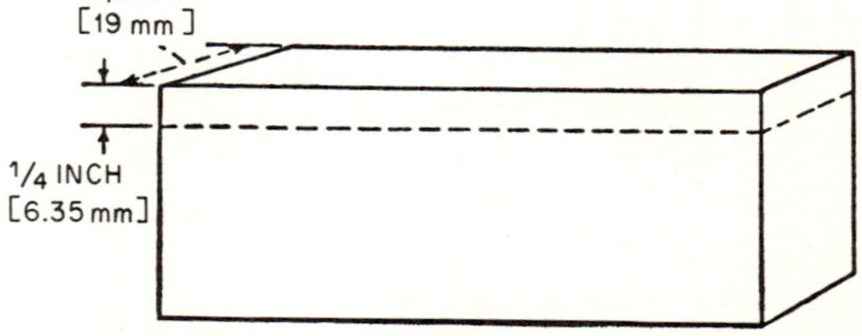

Fig. 21-6. Preparing the block for chiseling. Scribe one line around the block, as shown by the dashed lines.

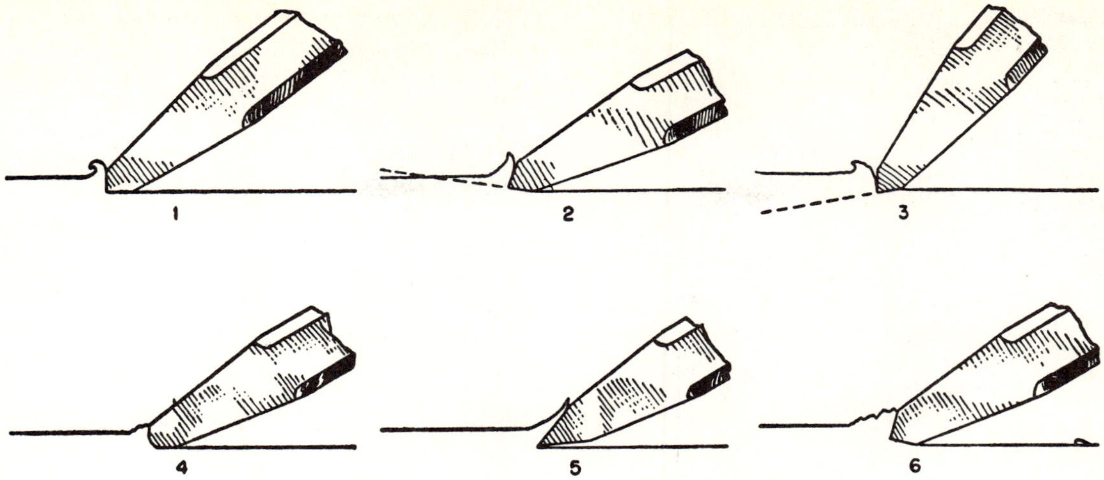

Fig. 21-7. Effects of improper chisel use. (1) Chisel held at the correct angle, with the lower face parallel to the surface being chiseled. (2) If the end of the chisel is held too low, the chisel will cut up out of the work. (3) If the chisel is held too high, the chisel will cut down into the work. (4) A dull chisel will work very hard and produce a rough, uneven cut. (5) A chisel ground at too great an angle will be awkward to handle and will soon dull. (6) A chisel ground at too blunt an angle will not cut well; the cut will be rough and hard to complete.

d. When chipping cast iron, do not chip all the way to the finishing (back) edge. This might cause a large piece to break off. Instead, stop when you approach the finishing edge. Start a second cut from the finishing edge back toward the first cut. Be careful to ease up on the hammer blows as the second cut nears the first, to avoid breaking off a piece of the cast iron.

_____ **11** Place a block of soft steel, 2 to 3 in [50 to 76 mm] wide, in the vise. Your job is to chip off the surface. To simplify the job, first use a cape chisel to cut one or two shallow grooves, parallel to each other, on the surface. Then use the flat chisel to remove the metal between the grooves.

Part Three CUTTING SHEET METAL

_____ **1** Place a piece of sheet metal in a vise. Make sure the part to be cut off projects above the jaws. Use soft jaws, as necessary, to protect the sheet metal from marring.

_____ **2** Use a flat chisel to shear off metal, as shown in Fig. 21-8. Start your cut at one edge, and move along the vise jaws. Hold the chisel at a 45° angle to the jaws to secure a good shearing cut.

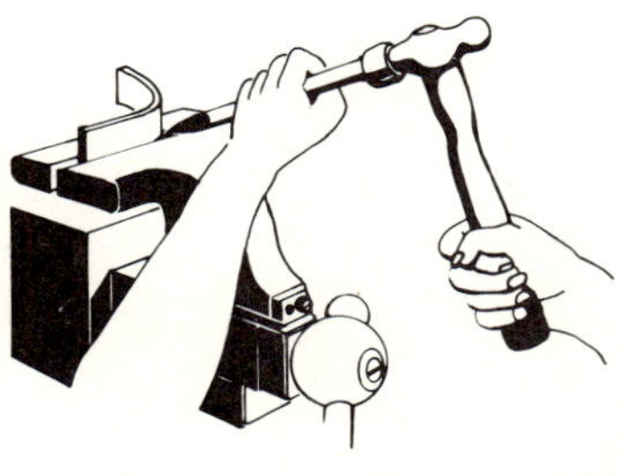

Fig. 21-8. How to shear metal clamped in a vise with a chisel and hammer. Note that the chisel is held at an angle.

Part Four CUTTING A KEYWAY IN ROUND STOCK

_____ **1** Use soft jaws to clamp a length of round stock in a vise. Place a wood block under the stock to hold it in position in the vise.

_____ **2** Use a steel rule and scriber to mark off, on the stock, the width and length of the keyway to be cut.

_____ **3** Select a cape chisel (Fig. 21-1) of the proper width for the cut. Hold the chisel at the proper angle, and begin cutting.

_____ **4** At the start, the chisel will have to be held so that it cuts down into the metal. Work from both ends toward the middle.

_____ **5** Continue cutting until the keyway is of the correct depth.

Part Five CUTTING RIVET HEADS WITH A CHISEL

_____ **1** Place a block, with peened-over pins or rivets, in a vise with the peened-over ends up. Place a piece of wood underneath the block.

_____ **2** Lay the face of the cutting edge of a flat chisel on the block, with the cutting edge at the peened-over (or riveted) end of the pin. Strike the chisel with a hammer to shear off the end of the pin.

_____ **3** For large rivet heads, first cut a groove down through the center of the head with a cape chisel. Then use a flat chisel to cut off the two halves.

Part Six CUTTING RODS WITH A CHISEL

_____ **1** Put a length of $\frac{1}{4}$-in [6.35-mm] soft steel rod vertically in the vise (Fig. 21-9).

_____ **2** Use a hammer and flat chisel to nick the rod deeply with two or three hammer blows (Fig. 21-9). Turn the rod 180°, and nick the other side. Then bend the rod back and forth until it breaks off.

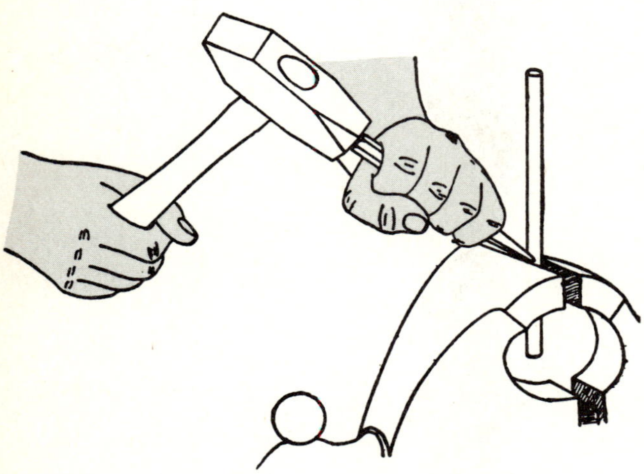

Fig. 21-9. A rod may be cut by nicking the rod on two sides with a chisel. Then bend the rod until it breaks.

True-or-False Test — Job 21

1 The flat chisel is used to cut keyways. **1** _____

2 If a flat chisel is cutting upward out of the metal, you should lower the chisel end to straighten the cut. **2** _____

3 To protect your eyes, you should wear goggles whenever someone is chipping with a chisel. **3** _____

4 If a flat chisel is cutting down too deeply into the metal, you should raise the chisel end to straighten the cut. **4** _____

5 It is dangerous to use a chisel with a mushroomed head. **5** _____

6 A flat chisel can be used to cut off rivet heads. **6** _____

7 Keyways can be cut with a cape chisel. **7** _____

8 Flying chips are dangerous to the eyes unless goggles are worn. **8** _____

9 Rivet heads should be cut off with a round-nose chisel. **9** _____

10 When chipping soft steel, you should make the cuts about $\frac{1}{4}$ in [6.35 mm] deep. **10** _____

Written Checkup — Job 21

1 What is the purpose of the cape chisel? The diamond-point chisel? The round-nose chisel?
2 What happens if the end of a flat chisel is raised when it is being used to chip a flat surface? If it is lowered?
3 Write down the procedure you followed in chipping a flat surface. Note any special details you found of interest.
4 Write down the procedure you followed in cutting sheet metal with a chisel.

NAME DATE CLASS

Objective: To learn how to use hacksaws.

Equipment: Various sections of pipe and round, flat, and rectangular stock of different metals.

Tools: Vise, hacksaw, hacksaw blades.

Job 22

HACKSAWS

PROCEDURE

_____ **1** Hacksaws are used for cutting metals (Fig. 22-1). They are, basically, a blade held in an adjustable metal frame (Fig. 22-2). The blades are replaceable.

_____ **2** Blades are made with from 14 to 32 teeth per inch. There is a proper blade, with the correct number of teeth, for each job. Using the wrong blade makes the job more difficult and could cause the blade to break. Figure 22-3 illustrates correct and incorrect blades for various jobs.

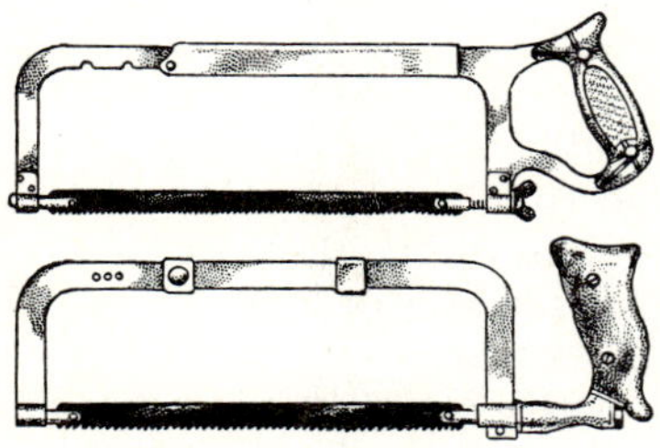

Fig. 22-1. Two types of hacksaws. (*General Motors Corporation*)

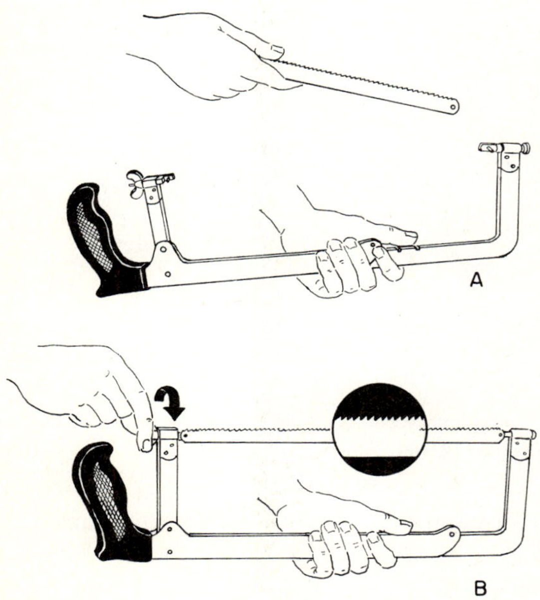

Fig. 22-2. (*A*) Installing the blade on a hacksaw. (*B*) Tightening the blade. Note that the teeth point away from the handle.

_____ **3** After the proper blade is selected, it should be placed in the hacksaw frame and tightened to the proper tension (Fig. 22-2). The teeth should point away from the handle, so that the cut is made on the push (forward) stroke.

_____ **4** When using the hacksaw (Fig. 22-4), make smooth, even, forward strokes. Use steady pressure to get a good cut. On the return stroke, lift the blade slightly so that the back edges of the teeth do not rub. Rubbing could dull the teeth.

_____ **5** In sawing sheet metal, you may find that, even with a fine blade, only one tooth is in contact with the work at any time. For better cutting action, clamp the sheet metal in the vise between two blocks of wood (Fig. 22-5). Then cut through the wood and metal at the same time.

_____ **6** Take good care of the blades. Do not throw them carelessly into a drawer with other tools, because this will dull the teeth. Instead, wipe them with a lightly oiled cloth. Then lay them flat in a drawer where they will not be damaged.

_____ **7** Clamp a piece of round stock in a vise. Select a blade with the proper number of teeth per inch to cut it (Fig. 22-3).

_____ **8** Insert the blade into a hacksaw frame. Tighten the frame by the means provided (Fig. 22-2). Be sure the teeth point away from the handle, so the blade will cut on the push stroke.

_____ **9** Hold the hacksaw as shown in Fig. 22-4. Grasp the handle firmly in one hand, and the end of the frame in the other.

_____ **10** Apply the blade to the stock at a selected point. Take a light first stroke to get the cutting started. The cut will start easier, and the saw will not slide off the selected point, if a starting nick is made with the edge of a file.

_____ **11** On the cutting stroke, move the hacksaw forward evenly and with uniform pressure. Take a full forward stroke. On the backstroke, lift the hacksaw slightly to avoid wear on the teeth. Avoid heavy pressure and twisting or kinking of the blade. Any of these may cause the blade to jam and break.

_____ **12** If you break a blade before a cut is complete, start the cut over again *at another place* with a new blade. The new blade will be a little wider (the

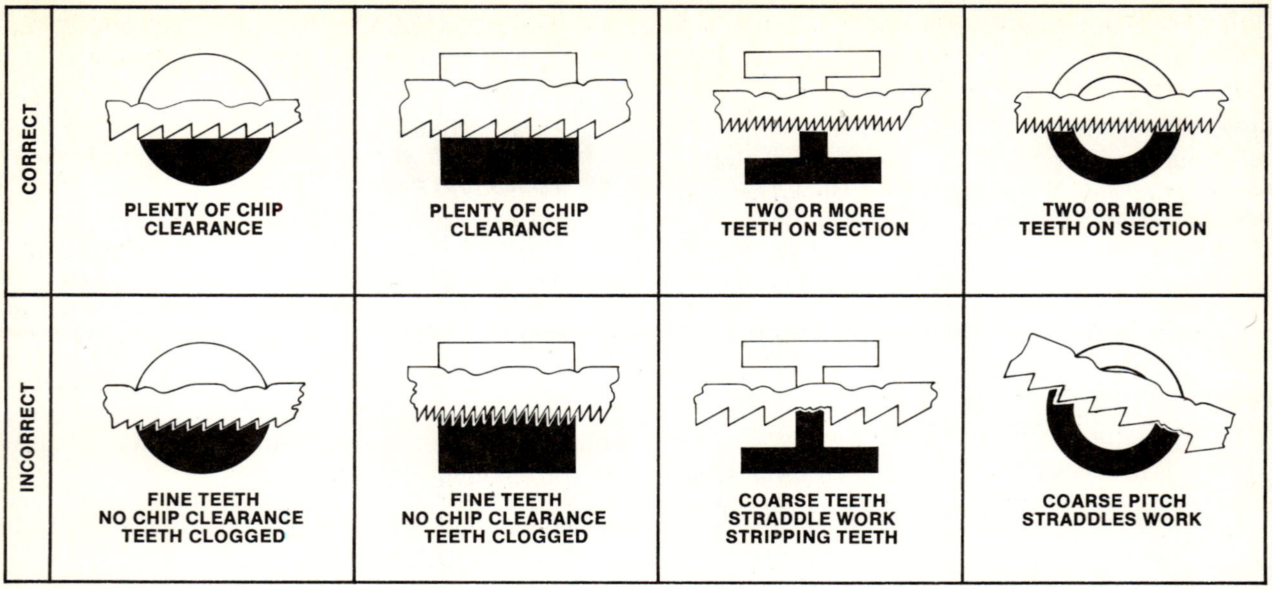

Fig. 22-3. Correct and incorrect number of teeth for various jobs.

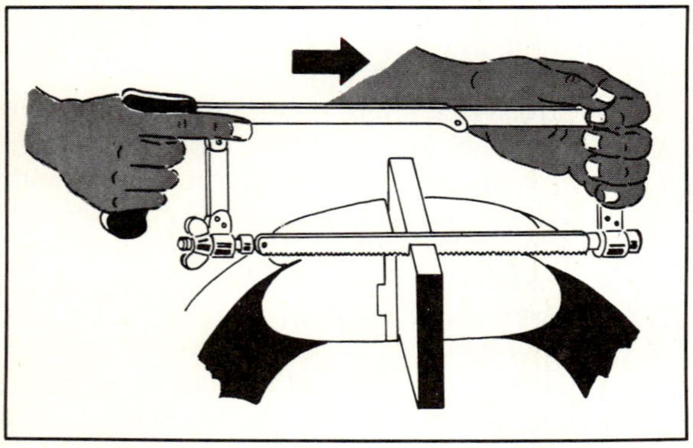

Fig. 22-4. How to hold and use a hacksaw.

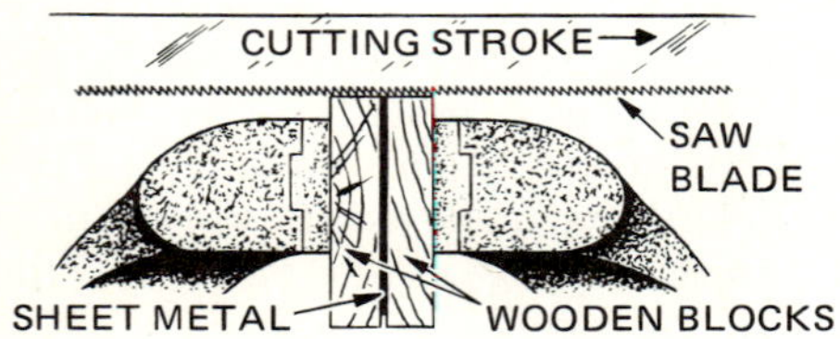

Fig. 22-5. To cut sheet metal with a hacksaw, clamp the sheet metal between two wood blocks in a vise. Saw through both wood and metal.

"set" of the teeth will not yet be worn). The greater width can cause the new blade to jam in the old cut. Starting a new cut with the new blade prevents jamming and breaking.

_____ **13** Complete the cut on the stock. Ease up on the pressure as you near the end of the cut. Often, in sawing pipe or other stock, it is not necessary to cut all the way through. Near the end of the cut, a slight pressure on the pipe or the stock will break it off at the cut. Burrs should be filed off (Job 23).

_____ **14** Try other cutting jobs with different materials and with blades of the proper coarseness. Note how the roughness of the cut depends on the coarseness of the blade.

_____ **15** Cut a piece of sheet metal. Clamp it between two pieces of wood in the vise. Cut through both pieces of wood and the sheet metal (Fig. 22-5). Note that, without the wood, it would be difficult to get through the thin metal.

_____ **16** Put two or three blades into the hacksaw frame, side by side. Cut a slot in the head of a bolt that is clamped in the soft jaws of the vise (Fig. 22-6). The more blades you use, the wider the slot you cut.

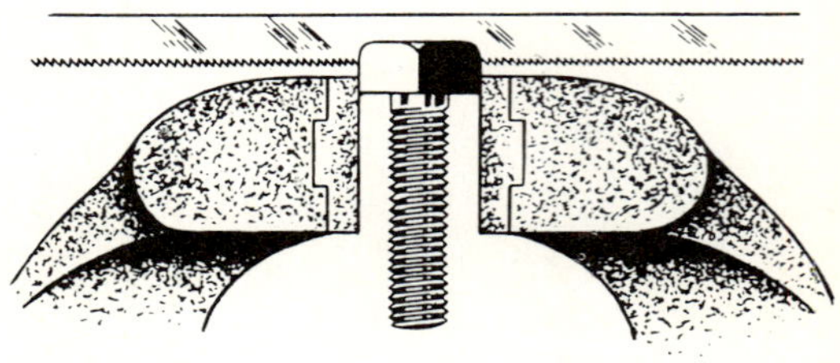

Fig. 22-6. Cutting a slot in the head of a bolt.

True-or-False Test **Job 22**

1 Only one blade can be put into a hacksaw at a time. **1** _______

2 The blade should be inserted in the frame so that it cuts on the push stroke. **2** _______

3 A coarse-toothed blade should be used to cut thin metal. **3** _______

4 A finer blade (more teeth per inch) should be used on tool steel than on soft steel. **4** ______

5 When a blade is broken before a cut is finished, a fresh cut should be started with the new blade.

5 ______

6 Heavy pressure on the blade is apt to break it.

6 ______

7 The cut will start more easily if a starting nick is made with a file. **7** ______

8 A new blade is narrower than a worn blade.

8 ______

9 The blade should not be lifted on the backstroke, since this causes wear of the teeth.

9 ______

10 Failure to tighten the blade sufficiently in the frame is apt to cause kinking of the blade.

10 ______

Written Checkup — Job 22

1 Write down the procedure for selecting a blade and sawing a piece of metal with a hacksaw.

2 Explain what would happen (and why) if a new blade were used to finish a cut started by an old blade.

NAME DATE CLASS

Objective: To learn about the types of files and their uses.

Equipment: Various sections of round and flat stock of different metals (some of these can be the pieces worked on in Jobs 19 and 21); stock with drilled holes in which pins have been inserted and peened over (Job 11); large bolts with battered threads.

Tools: Various files with handles, vise, file cleaner, chalk.

Job 23

FILES

Safety: *Be very careful that the handle is firmly in place on the file you are using. Never try to use a file without a handle. The tang is sharp enough to cut you severely if the file "hangs up" while you are filing. That is, if the file catches momentarily on the work, and stops moving, your hand will jam against the file tang. Never try to use the file as a pry bar; it is brittle and will break. Flying fragments of the file could cut your hands or face or give you a serious eye injury.*

PROCEDURE

_____ **1** A file is like a series of tiny chisels, each with a sharp cutting edge. Files come in many sizes and shapes, and have many uses. A typical file, with the parts named, is shown in Fig. 23-1.

_____ **2** As the face of the file is moved across a metal piece, the cutting edges or teeth remove shavings from the metal (Fig. 23-2). The coarseness of the file (Fig. 23-3) determines how thick the shavings are. The term *cut* is used to designate the coarseness or fineness of the file.

_____ **3** A coarse-cut file has relatively large teeth that are fairly far apart. A fine-cut file has many more teeth, much smaller and closer together. Such a file is called a *smooth,* or *dead-smooth,* file. Terms used to indicate coarseness or fineness are, in order: rough, coarse, bastard, second-cut, smooth-cut and dead-smooth.

_____ **4** A file with only one series of teeth across its face, in parallel lines, is known as a *single-cut* file (Fig. 23-4), regardless of its coarseness. A file with two series of teeth across its face in two different directions, is known as a *double-cut* file (Fig. 23-4).

_____ **5** In addition to their cut, files are classified according to shape. Files may be flat, triangular, half-round, or round. They may or may not be tapered (Fig. 23-5). The file selected depends on the job that is to be done.

_____ **6** There are a number of cautions to be observed in using a file. First, *never* attempt to use a file without a handle on the tang (Fig. 23-1). Otherwise, your hand might slip and the tang might be driven into your hand. To install a handle, put the

Fig. 23-1. Typical file and handle, with the parts named. (*General Motors Corporation*)

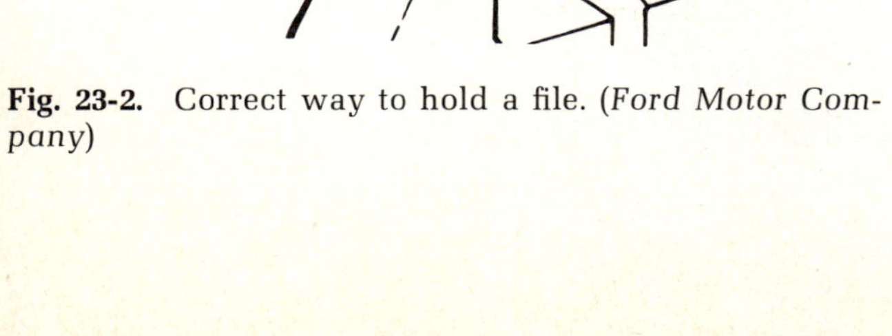

Fig. 23-2. Correct way to hold a file. (*Ford Motor Company*)

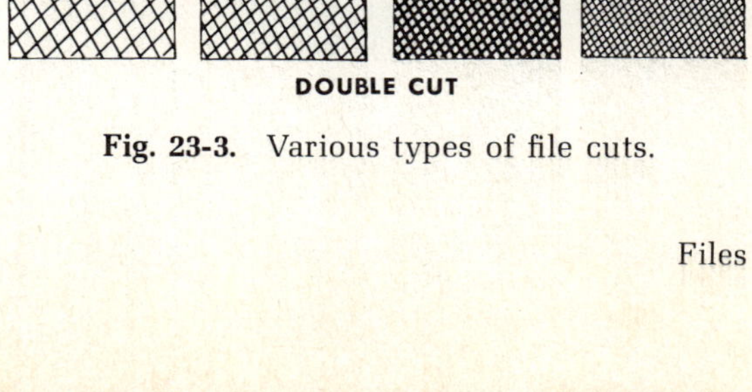

Fig. 23-3. Various types of file cuts.

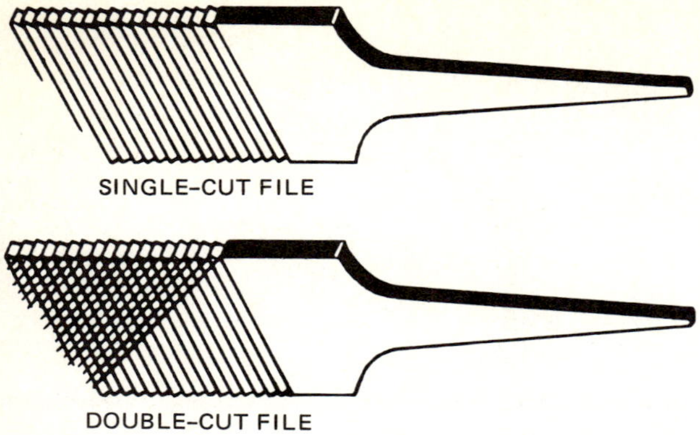

Fig. 23-4. Comparison of a single-cut file with a double-cut file.

file tang into the hole in the handle. Then tap the butt end of the handle on the bench (Fig. 23-6). This drives the file firmly into the handle. Never try to hammer the file into the handle. The file is brittle, and hammering could shatter it.

_____ **7** Be sure that the part to be filed is fastened down or clamped firmly in a vise. Use soft jaws in the vise if the part must be protected from marring.

_____ **8** In using the file, make the forward (cutting) stroke smooth and firm. Use the proper amount of downward pressure to get a cut (Fig. 23-2). A heavy pressure tends to overload the teeth so that the cut is uneven. A light pressure will not allow the teeth to cut properly, and may even dull the teeth.

_____ **9** On the back (return) stroke, lift the file clear of the work. Dragging the file teeth back over the work wears the cutting edges of the teeth. The exception to this is when the file is being used on soft metal, which tends to clog the teeth. In this case, dragging the teeth on the back stroke helps clean the teeth.

_____ **10** The teeth can also be cleaned with a file cleaner, or *file card* (Fig. 23-7). The file card is a wire-bristled brush that is moved across the face of the file in line with the teeth. Tapping the handle on the bench every few strokes also helps keep the teeth clean.

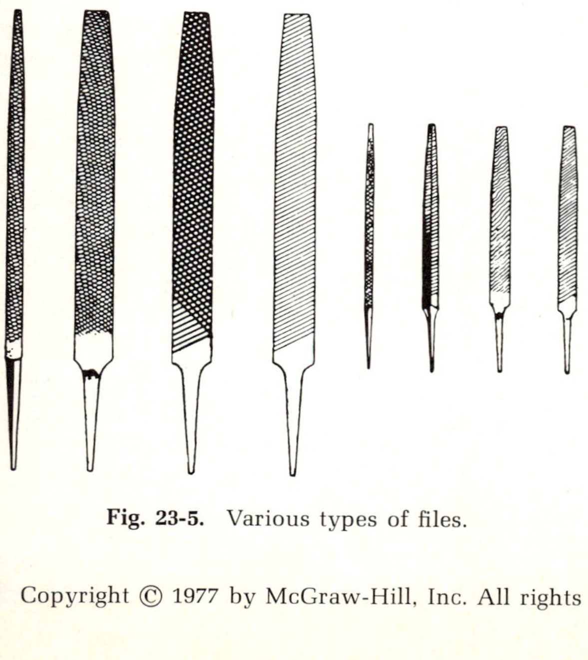

Fig. 23-5. Various types of files.

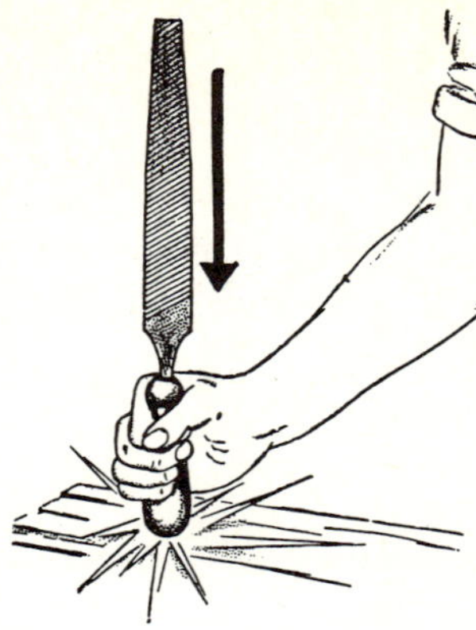

Fig. 23-6. Tapping the butt end of a file handle on the bench to tighten the file tang in the handle. (*General Motors Corporation*)

_____ **11** When a file is not in use, it should be put away carefully. If it is thrown into a drawer with other files or tools, the teeth will be chipped and dulled. Files are subject to rusting and should be kept in a dry place.

_____ **12** Never attempt to use a file as a pry bar, and never hammer on it. The file is brittle and could shatter dangerously.

_____ **13** Examine various files, and identify them. In the space below, write in the information about each file.

IDENTIFICATION AND CLASSIFICATION OF FILES

	LENGTH	CUT	COARSENESS	SHAPE
File 1	_____	_____	_____	_____
File 2	_____	_____	_____	_____
File 3	_____	_____	_____	_____
File 4	_____	_____	_____	_____
File 5	_____	_____	_____	_____

_____ **14** Clamp a piece of round stock in a vise, and prepare to file a flat on one end (Fig. 23-8).

_____ **15** Select a 12-in flat bastard file for the initial rough filing. Make sure the handle is firmly on the file tang; tap the handle end on the bench to make sure (Fig. 23-6).

_____ **16** Make sure the file is clean. Use a file card to clean it. Run the file card across the face of the file, so that the wire ends enter between the teeth and clean them out (Fig. 23-7). Do not try to clear the file by banging it on the table. This can break the file.

_____ **17** Take the handle of the file in one hand. Grasp the end of the file between the base of the thumb and the fingers of the other hand (Fig. 23-2).

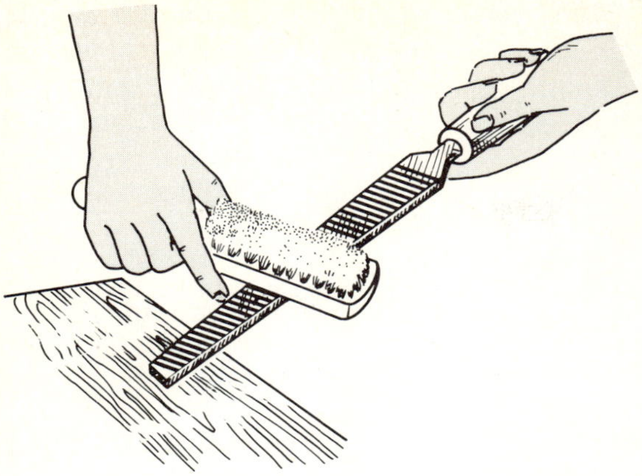

Fig. 23-7. Cleaning a file.

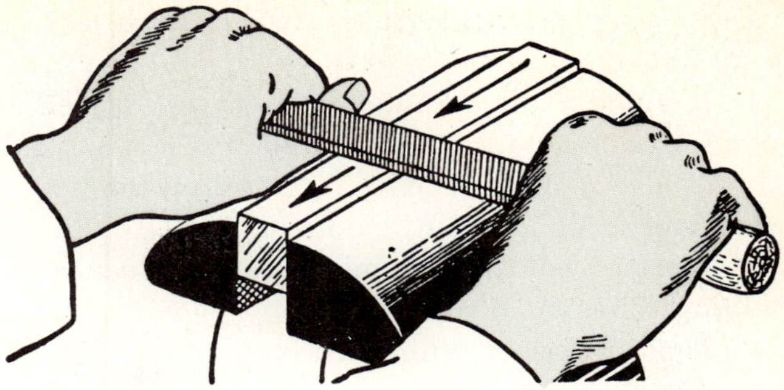

Fig. 23-9. Drawfiling. The file is held equally between the two hands and is drawn along the work in the direction shown by the arrows.

_____ **18** Lay the end of the file on the work clamped in the vise, and apply light pressure on the file. At the same time, push the file forward, in line with its length, so that it takes a cut. On the return stroke, lift the file clear of the work to avoid wearing the teeth. Experiment with different pressures, but avoid excessive pressure. A light pressure does not permit the teeth to take hold and cut. Heavy pressure causes too heavy a cut, so that the file teeth clog with chips.

_____ **19** Continue filing, noting how a flat is formed on the round stock. Avoid rocking the file, since this tends to round the flat. When the flat is deep enough (about halfway through the stock), lay aside the bastard file. Select a 12-in [about 30-cm] second-cut or smooth file to smooth the flat. Hold the file end between the thumb and first finger for finish work. Only a few strokes of the file should be necessary. See how smooth you can finish the flat.

_____ **20** Examine several of the pieces of stock you cut with the hacksaw in Job 22. Note that, where the cut finished, there are jagged edges of metal. These are called *burrs*. They almost always result from sawing, drilling, and threading metal. The burrs can be easily knocked off with a few strokes of a bastard or second-cut file. Select a burred piece, clamp it in the vise, and remove the burrs.

_____ **21** Select one of the pieces of flat stock that you chipped in Job 21 or that you cut with the hacksaw in Job 22. Clamp it in the vise, rough side up. Starting with a bastard or coarse file, and progressing to a smooth single-cut file, file the rough surface. At the same time, carefully note the actions of the different files. Work with even, steady strokes, and try to get the surface as smooth as possible. Don't rock the file, since this will round off the edges of the work. Clean the file from time to time with a file card. A clogged file does not cut as well as a clean file. It also tends to scratch the surface of the work.

_____ **22** Rub a piece of chalk on a second-cut or smooth single-cut file, so that the teeth are covered with chalk. Now try the file on the work. Note that it makes a smoother cut. The chalk acts as a lubricant, prevents tooth clogging, and helps the file produce a smoother surface.

_____ **23** Drawfiling is also used to finish a surface. In drawfiling, a single-cut file is grasped equally in the two hands. It is drawn across the surface at right angles to the usual file-stroke direction (Fig. 23-9).

_____ **24** Put a piece of stock with peened-over pins in drilled holes (from Job 11) into a vise. With a bastard file, carefully file off the head or peened-over part of a pin. See if you can remove the head without scratching the stock. After the head has been filed off, the pin is removed with taper and pin punches (Job 19).

_____ **25** Examine a bolt that has battered threads. Note how the threads are flattened, or pushed down, so that a nut will not turn on the bolt. Clamp the bolt head in a vise, with the battered threads up. Select a 6-in [about 15-cm] triangular file. With one edge, file into the root between two battered teeth. Use a rounding stroke to follow the circumference of the bolt. Continue to file until all the material that was battered down is removed, and the nut can be turned onto the bolt. If the job is well done, the nut should go on and turn easily.

NOTE: This is occasionally an important repair job. Cleaning up battered threads may save an expensive part (axle, armature, etc.). Although a thread chaser, or rethreader, is the preferred tool for such work, it is often not available. Then, the mechanic must rely on a file to clean up the threads.

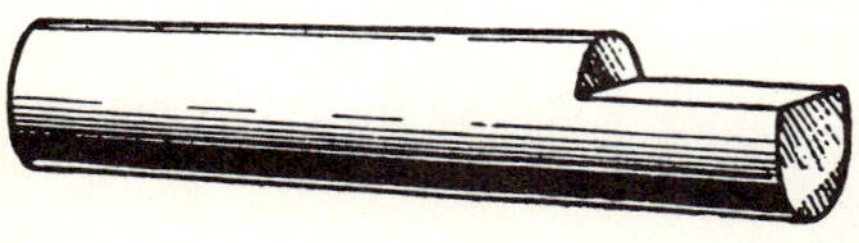

Fig. 23-8. Round stock with a flat filed on one end.

True-or-False Test — Job 23

1 A file contains many cutting edges. **1** _____

2 The coarser the file, the more cutting edges per inch. **2** _____

3 A heavy pressure causes the file teeth to clog with chips. **3** ______

4 In a double-cut file, the file teeth are formed in two series, extending in two directions. **4** ______

5 Chalk in a file acts as a lubricant so that a smoother filing job can be done. **5** ______

6 Battered bolt threads may often be cleaned up with a triangular file. **6** ______

7 The purpose of a file card is to file cards. **7** ______

8 A file should be cleared of chips every few strokes by striking it sharply against the workbench. **8** ______

9 A file should not be used without a handle. **9** ______

10 Double-cut files are used for fine finishing work. **10** ______

Written Checkup — Job 23

1 Write down the procedure for removing burrs and smoothing a rough surface with a file.

2 Write down the procedure for cleaning battered threads with a file.

3 Explain the difference between single- and double-cut files.

NAME DATE CLASS

Objective: To learn how to use taps.

Equipment: Pieces of metal stock with holes which can be topped.

Tools: Set of taps with tap wrench, electric and twist drills, vise.

Job 24

TAPS

PROCEDURE

_____ **1** Taps (Fig. 24-1) are used to cut screw threads in holes in metals and plastics. A tap looks somewhat like a screw, but the tap has a square end to fit into a tap wrench. *Flutes,* or grooves, run the length of the threads. They allow chips cut from the metal to escape, so they do not jam the tap threads (Figs. 24-1 and 24-3).

_____ **2** Taps are made in several styles (Fig. 24-1). To tap a hole that has been drilled, the square end of the tap fits into a tap wrench (Fig. 24-2).

_____ **3** Taps are very hard, as they must be to cut threads in metal. Therefore, taps are very brittle. The smaller the diameter of the tap, the easier it will break. Be very careful when using taps with a diameter of $\frac{1}{8}$ in [3.175 mm] or less. These small taps break easily, and a broken tap can be very difficult to remove. Sometimes, a part with a broken tap in it must be replaced with a new part, because it is impossible to remove the broken piece of the tap.

_____ **4** Taps are available in many different sizes. The tap size is the classification of the threads being cut. For example, to tap a hole for a $\frac{1}{4}$-20 bolt, you use a tap marked $\frac{1}{4}$-20. Most taps have their sizes etched on the shank.

_____ **5** To tap a hole that goes completely through a piece of metal (Fig. 24-3), you use a *taper* tap. To thread a hole only partway, you use a *plug* tap (Fig. 24-4). To make threads all the way to the bottom of a hole that is open on only one end, you use a *bottoming* tap (Fig. 24-4).

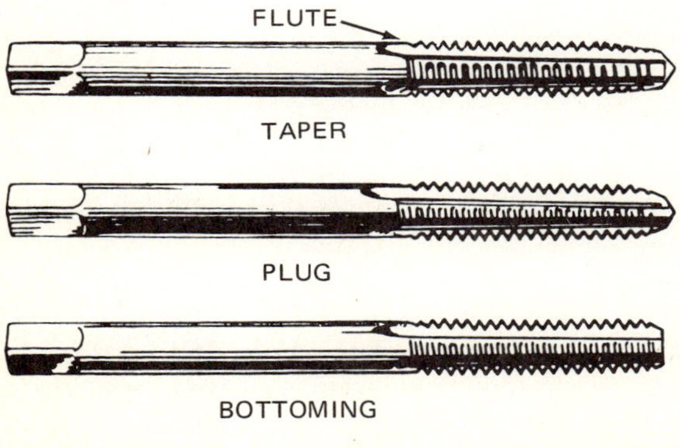

Fig. 24-1. Types of hand taps. (*Greenfield Tap and Die Corporation*)

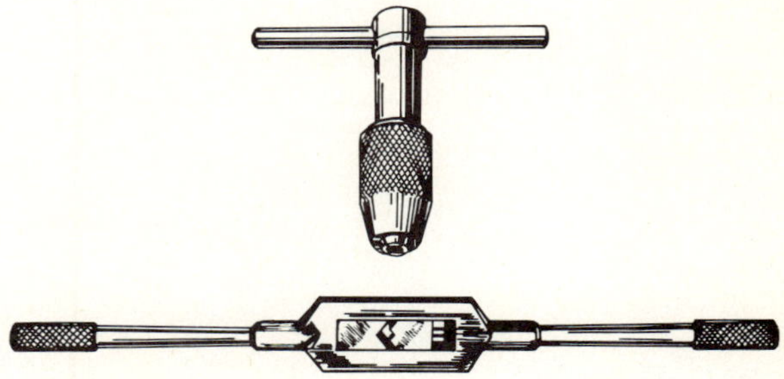

Fig. 24-2. Tap wrenches. (*Greenfield Tap and Die Corporation*)

_____ **6** To make a tapped hole in a piece of metal, first determine the thread size you need. Then consult a tap drill chart (Table 24-1). This chart shows you the size of drill required to make the proper hole for the threads. Selection of the tap is based on the size of the hole, the type of thread to be cut, and whether the hole is to be tapped all the way through the work.

_____ **7** Drill the hole, observing the proper safety cautions. Then, place the square end of the tap in a tap wrench (Fig. 24-2), and tighten it in place.

_____ **8** Here is how to tap a hole that is open on both ends (Fig. 24-3): First, apply a lubricant such as soluble oil to the tap. Then start the tap squarely in the hole (Fig. 24-3). This can be checked with a square. Then turn the tap wrench smoothly so that

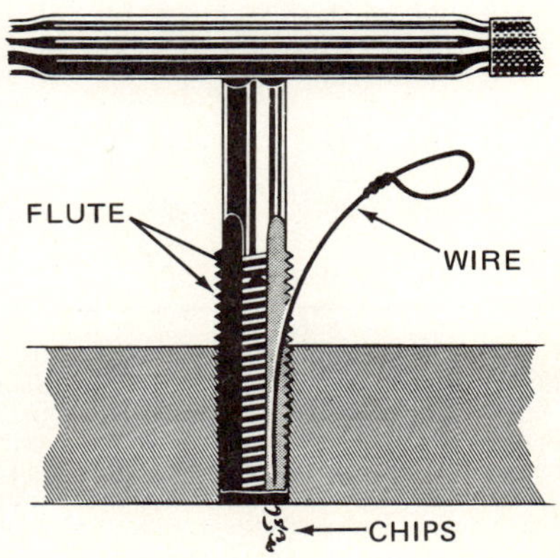

Fig. 24-3. Using a wire to clean metal chips from the flutes of a tap.

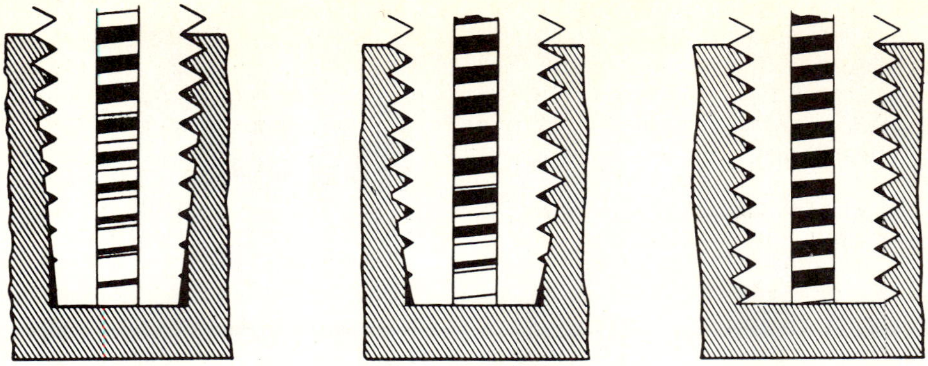

Fig. 24-4. Cross-sectional view showing the action of taper, plug, and bottoming taps in threading a blind hole.

TABLE 24-1

THREAD SIZE AND THREADS PER INCH	DRILL* SIZE	DRILL* NO.	THREAD SIZE AND THREADS PER INCH	DRILL* SIZE	DRILL* NO.
1–64	0.0595	53	10–32	0.1610	20
1–72	0.0595	53	12–24	0.1770	16
2–56	0.0670	82	12–28	0.1800	15
2–64	0.0700	50	1/4–20	0.1990	8
3–48	0.0781	5/64	1/4–24	0.2090	4
3–56	0.0810	46	1/4–32	0.2187	7/32
4–40	0.0890	43	5/16–18	0.2570	F
4–48	0.0935	42	5/16–24	0.2720	I
5–40	0.0995	39	5/16–32	0.2812	9/32
5–44	0.1040	37	3/8–16	0.3125	5/16
6–32	0.1065	36	3/8–24	0.3320	Q
6–40	0.1130	33	3/8–32	0.3437	11/32
8–32	0.1360	29	7/16–20	0.3906	25/64
8–36	0.1360	29	7/16–28	0.4162	13/32
10–24	0.1470	26	1/2–20	0.4531	29/64

*Drill sizes are designated in four ways—by numbers (Nos. 80 to 1, or from 0.135 to 0.228 in), by letters (A to Z, or from 0.234 to 0.413 in), by fractions (1/64 to 3 1/2 in), and by millimeters (in the metric system). The drill sizes indicated in the table are based on approximately 75 percent thread depth; that is, the tap will not give a full thread when it is run down into the hole but will give a 75 percent thread. This means that the top 25 percent, or one-fourth of each thread crest is absent. However, the remaining 75 percent is sufficient for most commercial purposes. For special precision applications, a fuller thread may be desirable. The fuller thread increases the thread strength only slightly (perhaps 5 percent) but makes the tapping job much harder; many more taps are broken.

the tap begins to cut threads in the metal. Advance the tap forward about one-half turn, then back one-quarter turn. This forward-and-back motion allows the metal chips to enter the flutes and drop out the bottom of the hole. When you tap holes in steel, be sure to use plenty of tapping compound of soluble oil to lubricate the threads.

_____ **9** As you will notice when you begin tapping a hole, you must press downward on the tap for the first few turns. But once the tap has made two or three turns, no more downward pressure is necessary.

_____ **10** You can tell how the tap is cutting by the feel of the tap wrench. The chips cut from the metal must not be allowed to fill the flutes. This makes turning difficult. It may even jam the tap in the hole, causing the tap to break. As soon as you feel any extra resistance, or a springiness, turn the tap backward about one-quarter turn to clean the flutes. Then add more lubricant and continue tapping (if the hole goes all the way through the workpiece). Otherwise, remove the tap from the hole, and clean and lubricate the tap.

_____ **11** A *blind* hole is a hole that does not go completely through the piece of metal. All three kinds of taps are needed (Fig. 24-4) to tap this type of hole. Begin with the taper tap. Then use the plug tap, which will go closer to the bottom of the hole. Finish the job with a bottoming tap, which should make threads to the bottom of the hole (Fig. 24-4).

_____ **12** When you are tapping a blind hole, you must be very careful to keep the tap clean and free of chips. Start the tap by turning it down one-quarter turn. Then back it one-eighth turn. Then move it forward one-half turn, and back one-fourth turn. About every two turns, back the tap out the hole and clean the chips from the tap. If possible, turn the work over, and shake out any chips that have collected at the bottom of the hole. If the work is too large to turn over, use a piece of wire to fish out the chips. Chips at the bottom of a blind hole prevent the taps from making threads all the way to the bottom of the hole.

_____ 13 After using a tap, brush off any chips, coat it with oil, and replace it in the tap set.

_____ 14 Drill and tap a hole in a metal block. Test the tapped hole with the appropriate screw. Start by determining the size of screw for which you will prepare the tapped hole.

_____ 15 Clamp the block in a vise. Drill a hole through it with the proper drill (see Table 24-1). For example, for a 10–32 screw, you would use a No. 20 drill.

_____ 16 Select the appropriate taper tap, and put it into the tap wrench. Guide the end of the tap into the hole with the fingers of one hand. At the same time, turn the tap wrench with other hand. Use enough pressure to force the tap to begin cutting, not ream out the top of the hole (Fig. 24-5).

_____ 17 Be sure the tap enters the hole squarely. If it starts in crooked, it will break unless it is removed and restarted. When restarting the tap, lean it slightly the other way. Do not bend it, since it is brittle and will break.

_____ 18 Once the tap is started, no further downward pressure is needed. The tap will continue to cut down into the hole as it is turned. Turn the wrench smoothly and evenly, without jerking it. Apply lubricant as required. Finish tapping the hole.

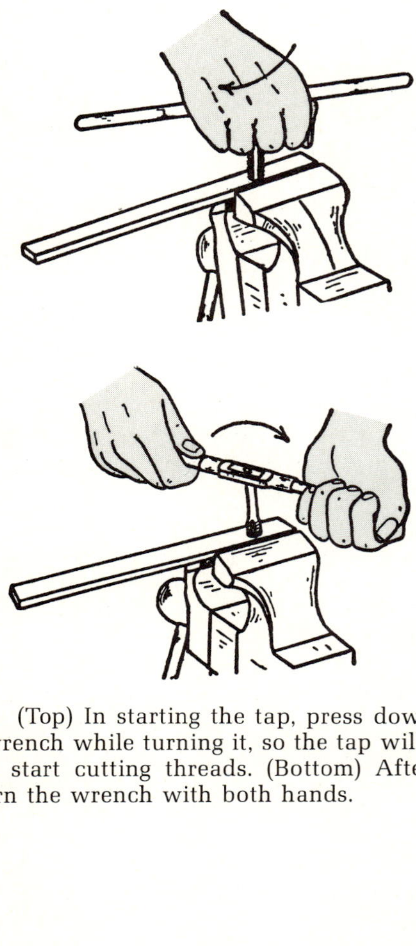

Fig. 24-5. (Top) In starting the tap, press downward on the tap wrench while turning it, so the tap will enter the hole and start cutting threads. (Bottom) After the tap starts, turn the wrench with both hands.

_____ 19 When such metals as steel and bronze are being tapped, the tap should be lubricated every turn or so with soluble oil. Cast iron and soft metals, such as brass, require no lubricant. Aluminum and aluminum alloys should be lubricated.

_____ 20 After the threads are cut, remove the tap and clean out the hole. Test the threads by turning a screw into the hole. The screw should turn in easily, but should not wobble when it is about halfway down.

_____ 21 Next, tap a large-diameter blind hole (a hole not drilled completely through). Use a taper tap first. Run it down most of the way into the hole. Then, use a plug tap to complete the thread cutting. If the threads must run all the way down into the hole, a bottoming tap should then be used. When you tap a blind hole, do not jam the end of the tap into the bottom of the hole. This might cause the tap to snap off. Use care when the tap gets near the bottom of the hole.

_____ 22 Drill and tap a small-diameter blind hole. Test it with a screw of the correct size and thread.

True-or-False Test — Job 24

1 For an 8–32 thread, you drill the hole with a $\frac{7}{32}$-in drill. 1 _____

2 In tapping a blind hole, the taper tap is used first, followed by the plug tap. 2 _____

3 When steel is being tapped, the tap should be lubricated. 3 _____

4 If the tap enters the hole crookedly, it can be rapped lightly with a hammer to straighten it. 4 _____

5 If the threads must run all the way down into a blind hole, the bottoming tap is used first, followed by the taper tap. 5 _____

Written Checkup — Job 24

Write down the procedure you followed in drilling and tapping a through hole. Then write down the procedure for drilling and tapping a blind hole.

Objective: To learn how to use thread dies.
Equipment: Rods of steel, brass, and other metals.
Tools: Set of thread dies, die stock, file.

Job 25
THREAD DIES

PROCEDURE

_____ **1** A die is run over the outside of a rod to make external, or outside, threads (Fig. 25-1). The die is held in a die stock (Fig. 25-2).

_____ **2** Remember, from Job 24, that taps are not adjustable. However, some method must be available to vary threads that are being cut, to provide a tight or loose fit. Many dies have an adjusting screw. The screw can be set to cut the standard thread depth, or to cut slightly oversize or undersize threads (Figs. 25-1 and 25-2). With the adjusting screw (Fig. 25-2), the die can be set to cut 0.001 to 0.003 in [0.03 to 0.08 mm] under or over the standard size.

_____ **3** Not all dies are alike. Some small dies are not adjustable (Fig. 25-1). Other dies are 12-sided (have 12 points) on the outside. These dies may be turned with a 1-in socket in tight places, where a die stock cannot be used.

_____ **4** Figure 25-2 shows an adjustable round die and its die stock. The adjusting screw is used to make a fine adjustment of the depth of thread to be cut. Tightening this screw spreads the die slightly. This causes the die teeth to make a slightly shallower cut, resulting in a tighter fit.

_____ **5** Another type of die shown in Fig. 25-1 is the round split die. This type of die is used in a die stock which has setscrews. The setscrews are tightened to make a fine adjustment of the thread depth.

_____ **6** As you can see in Fig. 25-2, there is a shoulder on the inside bore at the top of the die stock. The die is always placed in the die stock so that the chamfered teeth are away from the shoulder. Now examine the die closely. Note that there is a chamfer on one side of the die. The die teeth are cut straight to the back of the die. The chamfered opening is the side that you place on the rod to be threaded.

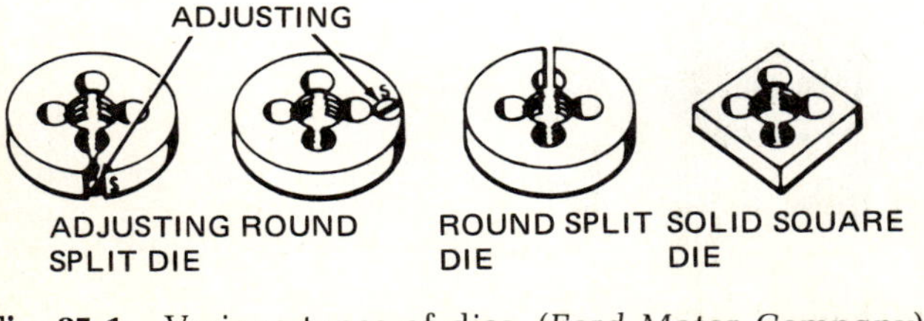

Fig. 25-1. Various types of dies. (*Ford Motor Company*)

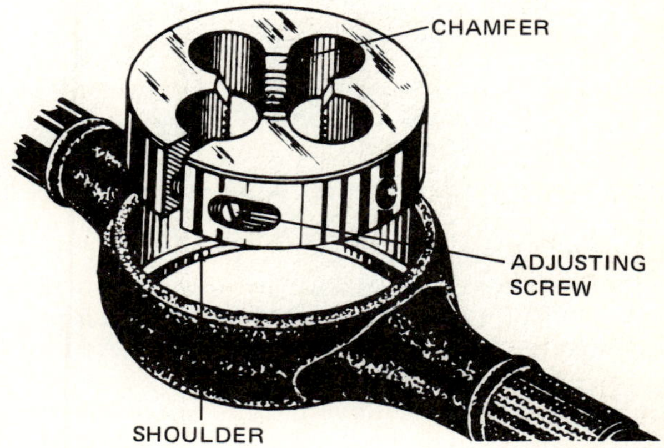

Fig. 25-2. Die stock with die above it, ready to be put into place.

_____ **7** Thread cutting with a die is similar to thread cutting with a tap (Job 24). The die is first fastened into the die stock (Fig. 25-3).

_____ **8** The rod to be threaded must be of the correct diameter. Threading an undersize rod results in a screw that is a sloppy fit in the tapped hole. The end of the rod should be cut square and then chamfered slightly, so that the die can get a start on the rod. The chamfer on the end of the rod helps to start the die squarely.

_____ **9** To thread a rod, brush some tapping compound or soluble oil on the rod. Then carefully start the die squarely with the work (Fig. 25-4). To start the thread, hold one handle in each hand, and

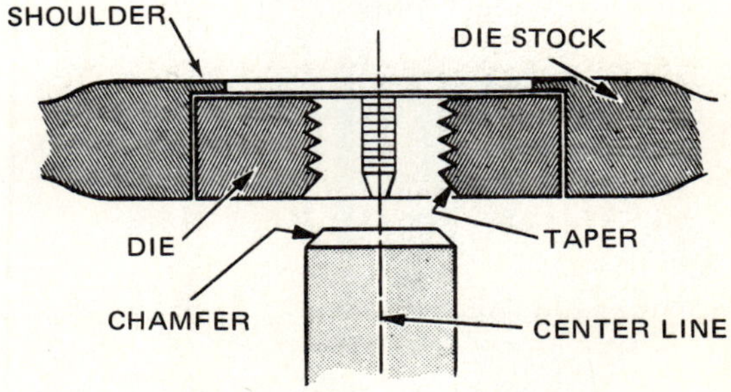

Fig. 25-3. Sectional view of a die stock, showing the chamfer on the threads of the die and the chamfer on the end of the rod to be threaded.

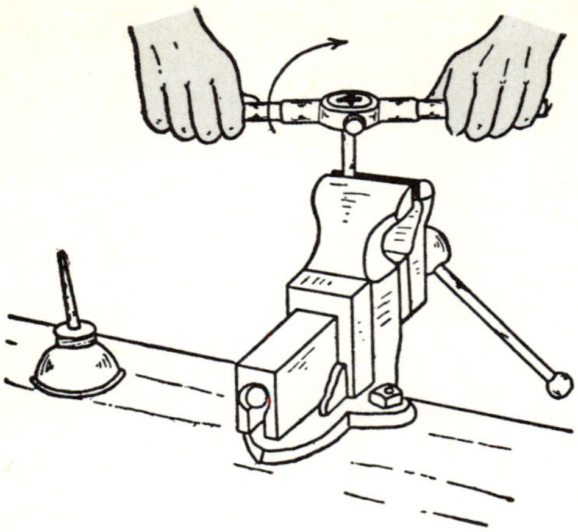

Fig. 25-4. To start the die, press down firmly on the stock while rotating it in the clockwise direction.

apply downward pressure while turning the die stock.

_____ **10** When you feel the die starting to cut teeth, turn the die stock forward one-half turn, and back $\frac{1}{4}$ turn (Fig. 25-5). This breaks off the chips and prevents them from jamming the threads. Add lubricant every couple of turns.

_____ **11** When you have cut the length of thread desired, remove the rod. Try the threads in the tapped hole.

_____ **12** As you will find out, cutting threads to the proper fit is a trial-and-error procedure. If the fit of the threads on the rod is too loose, cut off the threaded portion. Then expand the die by tightening the adjusting screw. The die will now cut shallower threads on the rod. This will produce a tighter fit.

_____ **13** If the fit is too tight, do not cut off the threads that you made. Instead, contract the die by loosening the adjusting screw. This decreases the size of the split in the middle of the die. The die will remove more metal, producing a looser fit. Run the die down the threads that were too tight. Then again test your threaded rod in the tapped hole.

_____ **14** When you are finished with the die and die stock, remove the die, clean it and oil it, and replace it in the set. Then clean and oil the die stock, and put it away carefully. This protects the die set from rust and damage.

_____ **15** Select a rod of an appropriate size, and place it in the soft jaws of a vise. Knock off any burrs on the end. Chamfer it slightly with a file, so that the die will start easily. Never try to thread hardened steel parts, since hardened steel is difficult to cut and may break the die.

_____ **16** Select a die of the proper size, and install it in the die stock. Never try to thread a rod with a die when the diameter of the rod exceeds the major diameter (outside diameter) of the thread. This will break the die or the rod.

_____ **17** Place the large opening of the die over the rod, and press down firmly on the stock. At the same time, start turning the stock in a clockwise direction (Fig. 25-4). The die should catch and start cutting a thread. If it does not, the rod is not of the proper size to fit the die, the die teeth are dull, or the die is upside down.

_____ **18** As soon as the die catches, lubricate it.

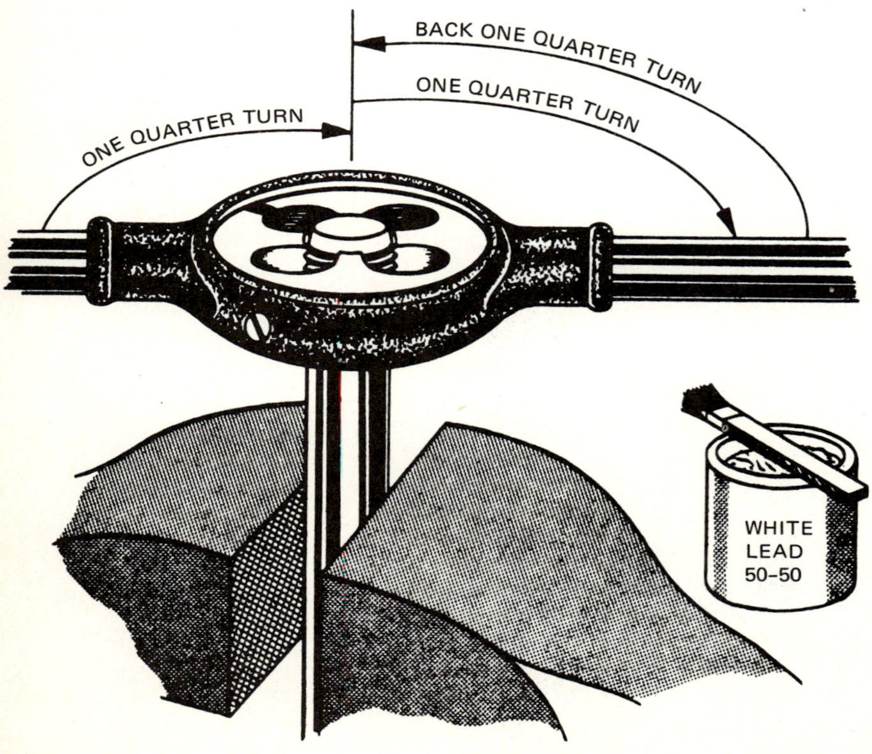

Fig. 25-5. To use a die properly, turn it ahead one-half turn, then back one-quarter turn, while keeping the threads coated with lubricant.

Then continue the threading job, moving one-half turn forward and then one-fourth turn back. Continue to add lubricant every couple of turns.

_____ **19** When the threads are finished, remove the die. Test the threads with a nut of the correct size. It should go on smoothly and evenly, and should not have any wobble, or play.

_____ **20** Cut additional threads, using rods of other sizes and other metals. Set adjustable dies to cut undersize and oversize threads. Then note the differences in thread fit.

True-or-False Test — Job 25

1 The die may break if it is used to thread a hardened steel rod. **1** _____

2 The die is turned in a clockwise direction to thread the rod. **2** _____

3 For proper fit, the diameter of the rod should exceed the major diameter of the thread. **3** _____

4 For every half turn forward, you should back the die off a full turn. **4** _____

5 The die should be tightened firmly in the die stock. **5** _____

Written Checkup — Job 25

1 Write down the procedure you followed in threading a rod.

2 Explain how to cut threads on a rod to fit very worn threads in a tapped hole.

NAME DATE CLASS

Objective: To learn how to remove broken studs or screws.

Equipment: Samples of metal stock that have been drilled and tapped, bolts or screws to fit tapped holes.

Tools: Stud extractors, wrench, electric drill, twist drills, hammer, center punch, hacksaw, file, diamond-point chisel, vise.

Job 26

STUD REMOVERS

Safety: *Observe the safety precautions regarding the use of electric drills, as outlined in Job 29.*

PROCEDURE

_____ **1** Sometimes a stud or bolt will break off (Fig. 26-1). Often this occurs while you are screwing a bolt into a hole. The bolt strikes chips or dirt at the bottom of the hole, and the head twists off. This can happen easily to smaller sizes of bolts. The bolt may snap off in two different ways. It may break with a fairly long piece sticking up above the hole ($\frac{1}{4}$ in [6.35 mm] or more), or break off flush with the hole. How the broken piece is removed depends on where the bolt broke.

_____ **2** When removing a broken bolt or stud, keep the threads covered with penetrating oil. If the threads are rusty, allow the penetrating oil to soak in for as long as possible, preferably overnight, before removing the broken piece.

_____ **3** Often, when a small bolt breaks, it is because the end of the bolt has bottomed. That is, the bolt may be too long, so that it strikes the bottom of the hole. Or, the hole may be partially filled with debris, such as chips and dirt. In either case, further turning of the bolt head twists it off. When this happens, the part that remains in the threaded hole is often fairly loose. It can be turned out easily once you get a good grip on it.

_____ **4** To remove a broken bolt sticking up above the hole, attach Vise-Grip or locking pliers to the protruding part (Fig. 26-1). Then try to turn the broken bolt. In most cases, the broken part can be turned out easily.

_____ **5** There are other ways to remove a broken bolt. If the break is above the surface, it may be possible to file flats on two sides (Job 23). Then a wrench can be used to back it out. Or, a slot may be cut so that a screwdriver can be used (Job 22). Light tapping with a hammer on the top and around the sides of a broken bolt may help loosen it.

_____ **6** A different procedure is required when the bolt breaks off about flush with the hole. You may have to drill out the center of the bolt, and remove the remainder with a screw extractor. However, the first and easiest way is to try removing the broken bolt with a center punch (Fig. 26-2). Place the point of the punch on the bolt, but not at its center. Tap the punch lightly with a hammer, to try to back the broken part out of the hole. This method will not work on bolts that break because of rust or cross-threading.

_____ **7** If the break is about level with or below the surface of the hole, a screw extractor (Fig. 26-3) is required. One type, called the Ezy-Out, is tapered and has a coarse spiral thread (Fig. 26-3). Other types of screw extractor have gradually tapering flutes (Fig. 26-4) instead of spiral threads. Both work by wedging against the shell of the broken bolt. Screw

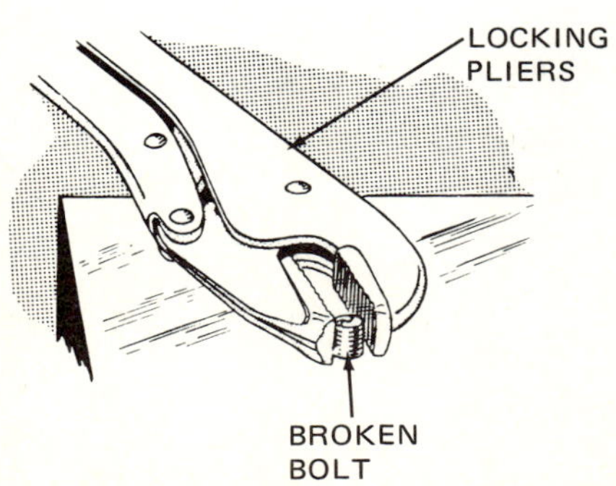

Fig. 26-1. Removing a broken bolt with locking pliers.

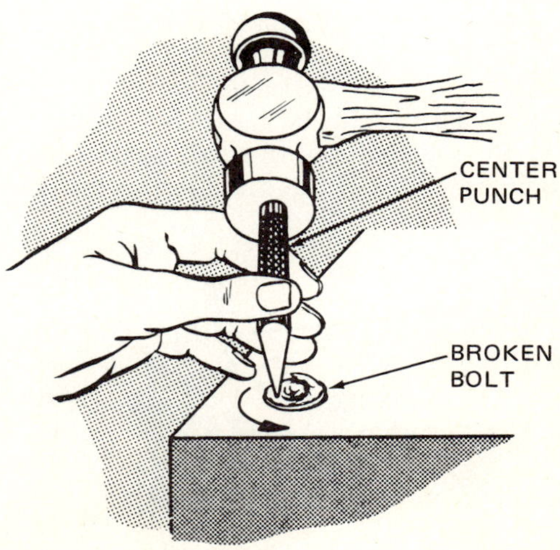

Fig. 26-2. Removing a broken bolt with a center punch.

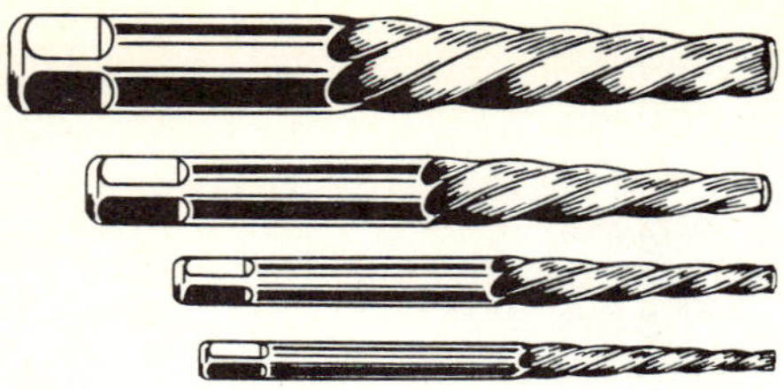

Fig. 26-3. Set of stud and bolt extractors.

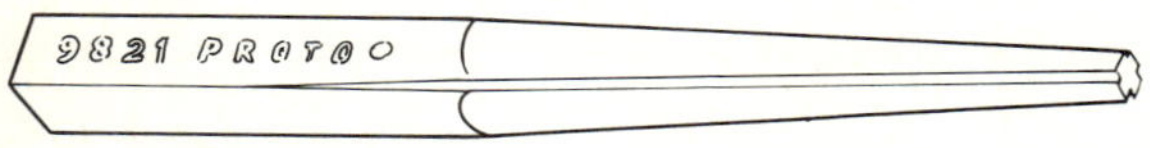

Fig. 26-4. Screw extractor which may be turned in either direction. (*Proto Tool Company*)

extractors are available in sets, because larger bolts require larger screw extractors.

_____ **8** The first step in removing a bolt broken off at the surface is to file the broken portion smooth, if possible (Job 23). Then center-punch the exact center of the bolt.

_____ **9** Next, determine the size of broken bolt you are trying to remove. Then, locate that bolt size in Table 26-1. In the right column of Table 26-1, across from the bolt size, is the drill size to use. Following good safety procedures, and wearing eye protection, drill out the center of the bolt (Fig. 26-5).

_____ **10** If the bolt has a fairly large diameter, use a small drill first. Follow this with a larger drill. Try to make a hole nearly as large as the small diameter of the threads (Fig. 26-6). This will leave only a thin shell.

_____ **11** Now, insert a screw extractor of the correct size into the hole. Drive the screw extractor in place with a few taps from a hammer (Fig. 26-7). This will cause the edges of the extractor to bite into the sides of the shell. To get the broken piece out, fit the tap handle from your tap-and-die set to the extractor. Turn the tap handle, using both hands, to turn out the broken bolt (Fig. 26-8). If a tap handle is not available, turn the extractor with a wrench (Fig. 26-9).

_____ **12** When no extractor is available, a diamond-point chisel can be used to remove a broken bolt (Fig. 26-10). Drill out the bolt, as outlined above. Then drive the end of the diamond-point chisel into

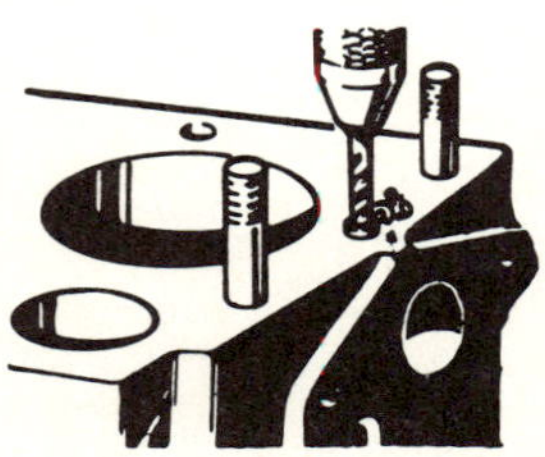

Fig. 26-5. Drilling a hole in a broken bolt. (*Proto Tool Company*)

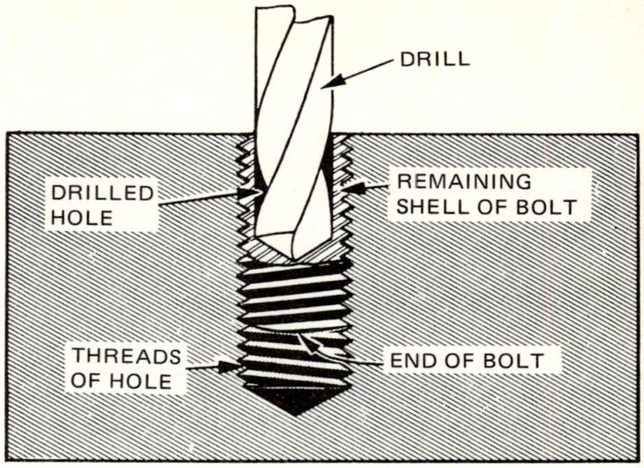

Fig. 26-6. Sectional view showing how a drill is used to take out the center of a broken bolt so that only a thin shell remains in the hole.

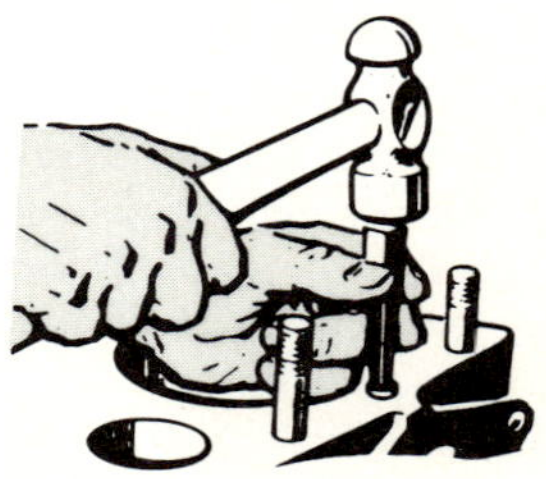

Fig. 26-7. Driving a screw extractor into the hole drilled in a broken bolt. (*Proto Tool Company*)

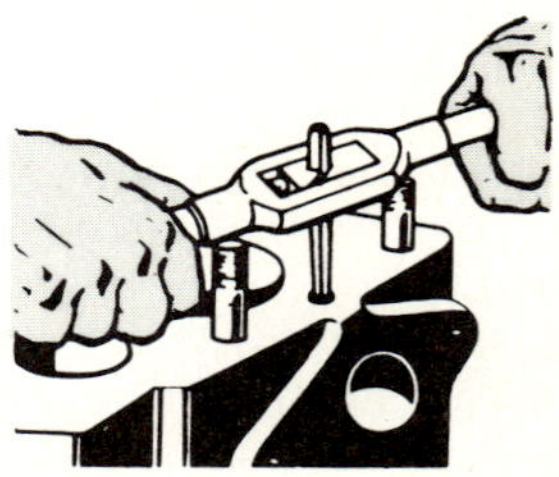

Fig. 26-8. Using a tap handle on the extractor to turn out the broken bolt. (*Proto Tool Company*)

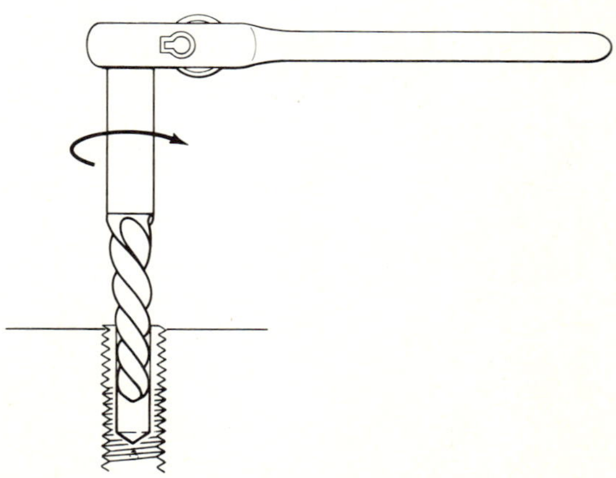

Fig. 26-9. Using a stud extractor to remove a broken stud. (*General Motors Corporation*)

TABLE 26-1

EXTRACTOR		USED FOR		
SIZE, NO.	OVERALL LENGTH, IN	NOMINAL SCREW AND BOLT SIZE, IN	NOMINAL PIPE SIZE, IN	USE DRILL SIZE DIAMETER, IN
1	2	$\frac{3}{16}$-$\frac{1}{4}$		$\frac{5}{64}$
2	$2\frac{3}{8}$	$\frac{1}{4}$ -$\frac{5}{16}$		$\frac{7}{64}$
3	$2\frac{11}{16}$	$\frac{5}{16}$-$\frac{7}{16}$		$\frac{5}{32}$
4	3	$\frac{7}{16}$-$\frac{9}{16}$		$\frac{1}{4}$
5	$3\frac{3}{8}$	$\frac{9}{16}$-$\frac{3}{4}$	$\frac{1}{4}$	$\frac{17}{64}$
6	$3\frac{3}{4}$	$\frac{3}{4}$ -1	$\frac{3}{8}$	$\frac{13}{32}$
7	$4\frac{1}{8}$	1 -$1\frac{3}{8}$	$\frac{1}{2}$	$\frac{17}{32}$
8	$4\frac{3}{8}$	$1\frac{3}{8}$ -$1\frac{3}{4}$	$\frac{3}{4}$	$\frac{13}{16}$
9	$4\frac{5}{8}$	$1\frac{3}{4}$ -$2\frac{1}{8}$	1	$1\frac{1}{16}$
10	5	$2\frac{1}{8}$ -$2\frac{1}{2}$	$1\frac{1}{4}$	$1\frac{5}{16}$
11	$5\frac{5}{8}$	$2\frac{1}{2}$ -3	$1\frac{1}{2}$	$1\frac{9}{16}$
12	$6\frac{1}{4}$	3 -$3\frac{1}{2}$	2	$1\frac{15}{16}$

the hole. Use a wrench to turn the chisel and remove the bolt.

_____ **13** Another method of removing a broken bolt or stud is to drill out the center of the broken piece as large as possible (Fig. 26-6). Be careful that the drill does not touch the inside of the threads. Then use a round-nose chisel to collapse the remaining shell (Fig. 26-11). Once collapsed, the shell of the bolt can usually be worked out easily.

_____ **14** Clamp a piece of metal stock in a vise. Turn some bolts into tapped holes in the stock. To make the bolts fit tightly, batter the threads lightly with a hammer before installing them.

_____ **15** Use a hacksaw to saw off the bolt heads. Saw some off flush with the surface of the stock. On others, leave $\frac{1}{4}$ to $\frac{1}{2}$ in [6.35 to 12.7 mm] of bolt protruding above the surface.

_____ **16** First select a bolt that protrudes above

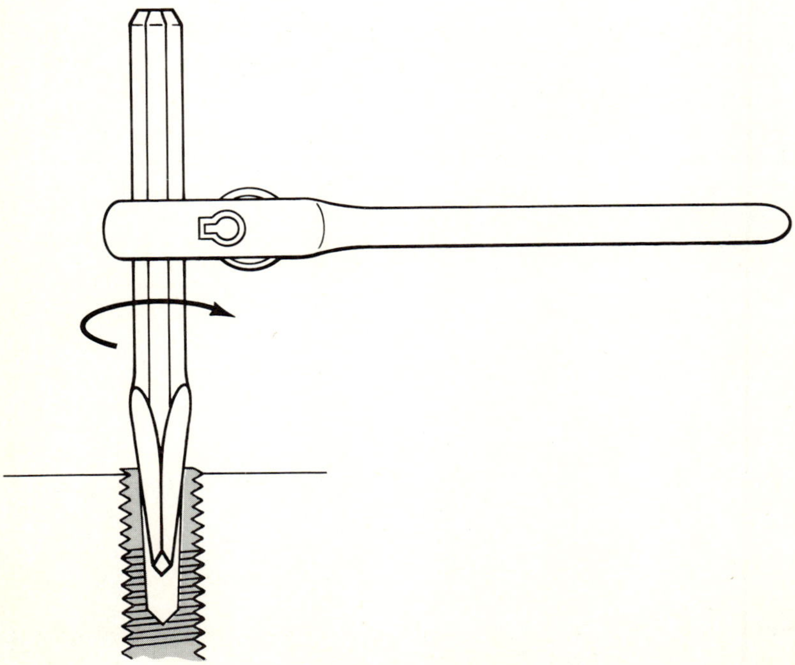

Fig. 26-10. Using a diamond-point chisel as an extractor to remove a broken bolt.

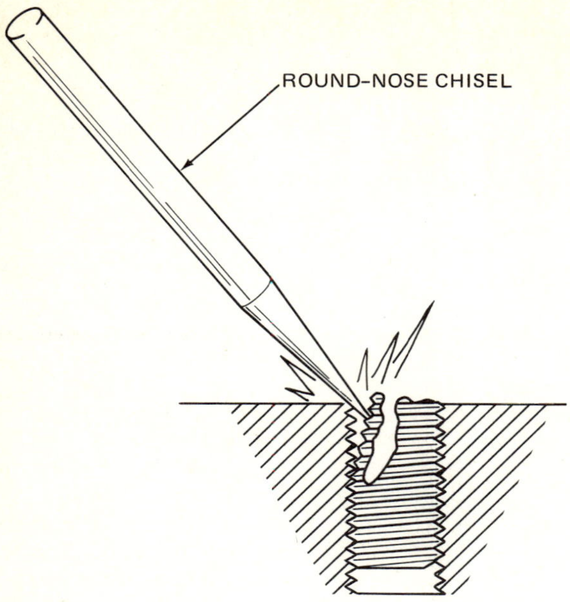

Fig. 26-11. Removing the drilled shell of a broken bolt by collapsing it with a round-nose chisel. (*Ford Motor Company*)

the surface. Use a file to file two flats on the bolt. Then use an adjustable wrench to back the bolt out.

_____ **17** Now select another bolt that protrudes above the surface. Use a hacksaw to cut a slot in the end of the bolt. Then use a screwdriver to back the bolt out.

_____ **18** Select a bolt that has been sawed off flush with the surface. Use a center punch to mark the center of the bolt end.

_____ **19** With an electric drill and a small twist drill, make a hole in bolt. Do not drill clear through bolt.

_____ **20** Choose a twist drill that is nearly as large as the small diameter (root diameter) of the threads. Enlarge the small hole by drilling the bolt with the larger drill.

_____ **21** Use a stud extractor to remove the bolt. The type of extractor with coarse threads is started into the hole and turned with a wrench until the bolt is backed out (Fig. 26-9). The fluted type is driven in with a hammer. The bolt is backed out by turning the extractor with a tap handle (Fig. 26-8).

_____ **22** If bolt is badly rusted into place, as sometimes happens, put a little penetrating oil around the bolt. Wait for several minutes. The penetrating oil often loosens up the threads so that the bolt comes out more easily.

_____ **23** If the extractor splits the bolt so that it expands and jams more tightly into the hole, remove the extractor. Collapse the shell of the bolt with a diamond-point chisel, and remove it (Fig. 26-11). Use care to avoid damaging the threads of the tapped hole. After removing the bolt, examine the tapped hole. If the threads are damaged, retap them with the proper tap.

True-or-False Test — Job 26

1 The stud extractor with the coarse spiral thread is turned in a clockwise direction to remove a broken stud. **1** _______

2 To use an extractor, you first drill a hole in the broken stud. **2** _______

3 Where the stud or screw end is far enough above the surface, a hacksaw can be used to cut a screwdriver slot in it. **3** _______

4 If the extractor splits the bolt, you should use a larger wrench to gain sufficient torque to back the bolt out. **4** _______

5 If a bolt is rusted in place, you should strike the extractor with a hammer to break the bolt loose. **5** _______

Written Checkup — Job 26

Write down the procedure you followed in removing studs or broken bolts with an extractor.

NAME DATE CLASS

Objective: To learn how to use the tubing cutter and flaring tool.

Equipment: Piece of copper tubing 12 in [304.8 mm] long, piece of steel tubing 12 in [304.8 mm] long.

Tools: Tubing cutter, double-flaring-tool set.

Job 27

TUBING CUTTER AND FLARING TOOL

PROCEDURE

_____ **1** Many systems on the automobile use copper or steel tubing connected with special fittings (Fig. 27-1). Any time a section of tubing is replaced, you probably will have to cut, bend, and flare it.

_____ **2** Rolls of tubing are available from dealership parts departments and local auto-parts stores. Steel-wire armor wrapping is also available; it is used to cover tubing that is exposed to road stone damage.

_____ **3** According to General Motors, copper tubing should not be used to replace brake lines, fuel lines, or oil-cooler lines on automobiles. Normal driving vibrations cause copper tubing to crack much sooner than steel tubing. The copper cracks because it gets brittle as it flexes with the vibrations. In addition, soft copper tubing does not resist scrapes, dents, crush, or road stone damage as well as steel tubing.

_____ **4** All fittings on steel lines must be double-flared. Seamless steel tubing can be bent and flared with a tubing cutter and a flaring tool. With the proper tools and a little practice, steel tubing is as easy to work as copper tubing.

_____ **5** The proper way to cut tubing is with a tubing cutter. Never use shears or a hacksaw, or break the tubing by bending it back and forth. A tubing cutter (Fig. 27-2) is similar in construction and operation to a pipe cutter. A cutting wheel and two rollers are positioned to hold the piece of tubing between them. The rollers are adjustable; they are moved toward or away from the cutting wheel by the adjusting screw.

_____ **6** When you have to cut tubing, place it in the tubing cutter (Fig. 27-2). Tighten the adjusting screw until the cutting wheel and rollers just touch the tube. The cutting wheel must touch the exact line where the tubing is to be cut. Then, turn the tubing cutter around the tubing while you gradually tighten the adjusting screw, a little at a time (Fig. 27-3). This prevents flattening the tube. Continue this procedure until the tubing is cut off. Do not tighten the adjusting screw too much at a time, or the rollers and cutting wheel will crush or split the tubing.

_____ **7** When the tubing is cut, a burr is left on the inside of the tubing. The burr must be removed. Figure 27-4 shows how to remove the burr with the reamer on the back of the tubing cutter. After deburring, any small pieces of loose metal should be wiped or blown from the inside of the tubing.

_____ **8** The only safe way to bend copper, aluminum, and steel tubing is with a tubing bender. Several different types are available. Figure 27-5 shows one type. It will bend the most comonly used sizes of tubing at angles of up to 180°. This tubing bender has a movable forming block with several tubing channels. Each channel accepts one of the four different sizes of commonly used automotive tubing.

_____ **9** To bend a tube, back off the adjusting screw. The forming block will move away from the pressure posts of the bender (to the right in Fig. 27-5). This opens the tubing bender. Insert the tubing, through the bender, in the correct size channel. Then tighten the adjusting screw with a ratchet wrench (Fig. 27-5). This forces the forming block to move toward the posts and bend the tubing. For heavy jobs, the bender can be mounted in a vise. In general, you should never try to bend tubing without a tube bender.

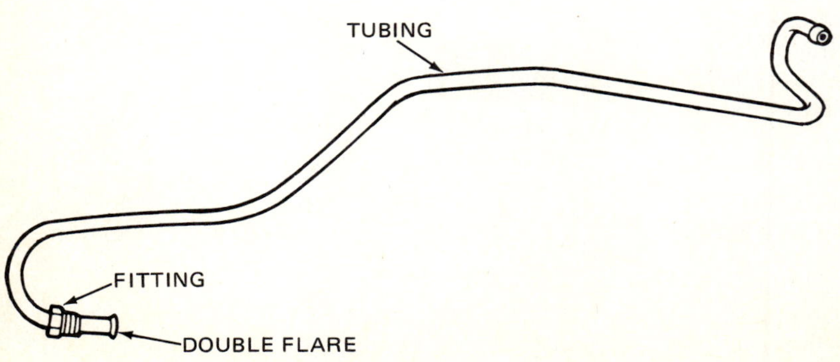

Fig. 27-1. Fuel line. (*Ford Motor Company*)

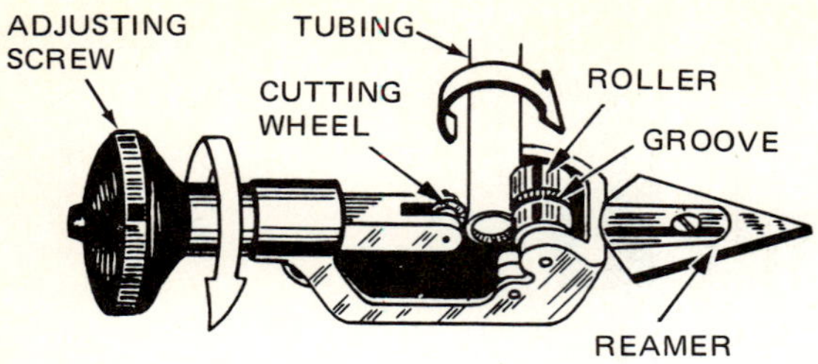

Fig. 27-2. Cutting tubing with a tubing cutter. (*Imperial Brass Mfg. Co.*)

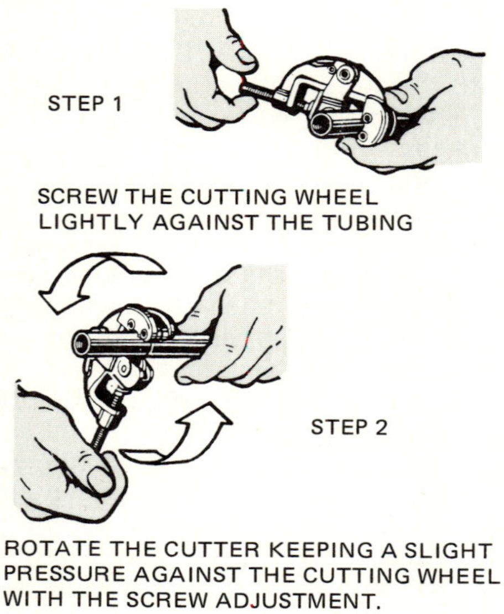

Fig. 27-3. How to use a tube cutter. (*Imperial Brass Mfg. Co.*)

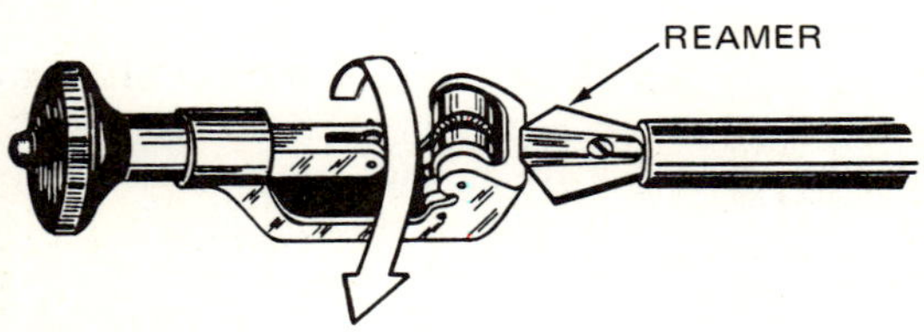

Fig. 27-4. Using the reamer on the tube cutter to remove the burr from the inside of the tubing. (*Imperial Brass Mfg. Co.*)

_____ **10** Tubing is used to connect the parts of air- or liquid-operated systems. To prevent leakage at the connections, the ends of the tubing are flared. A flare is a widening at the end of the tube. It is shaped like a funnel and held in the connector by a threaded fitting. A flaring tool (Fig. 27-6) must be used to flare a tube without cracking it.

_____ **11** There are two types of flare, the single flare and the double flare. Automotive connections require double flares. Single flares are not strong enough, and may crack in service.

_____ **12** A fitting or flare nut is often used to fasten a flared tube to a flare fitting. If so, the fitting or nut should be placed on the tube before it is flared.

_____ **13** As you can see in Fig. 27-6, the flaring tool has several different sizes of openings. There is one opening for each size of tubing that the tool can flare. If necessary, square the ends of the tubing with a file (Job 23). Then ream the tubing again (Fig. 27-4). Make certain that no chips or filings are left in the tube.

_____ **14** Loosen the wing nut of the flaring tool until the tubing will fit in the proper opening (Fig. 27-6). Place the tube in the opening, with the end slightly above the tool. Then, holding the tubing in place, tighten the wing nut (Fig. 27-6).

_____ **15** Select the correct size of double-flare adapter from the flaring kit. Place it in the end of the tube that is sticking up above the flaring tool (Fig. 27-7). Then place the yoke flaring head in the recess in the adapter. Tighten the handle of the yoke (Fig. 27-6). Figure 27-7 shows this first step in the double-flaring procedure.

_____ **16** Loosen the yoke, remove the adapter, and note the condition of the flare. It should be as shown in Fig. 27-7. The first flaring operation should leave the end of the tubing bent outward and upward at about a 45° angle.

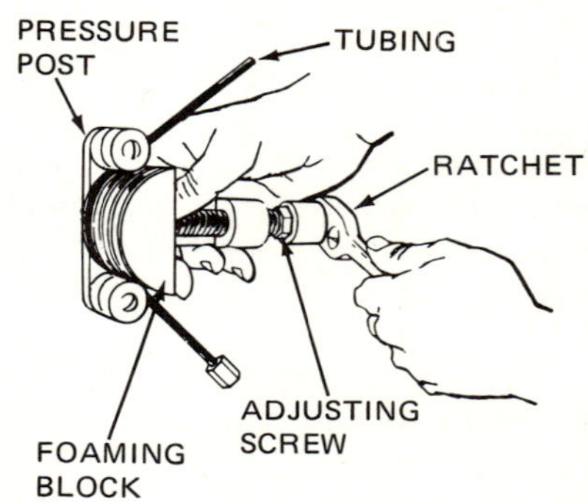

Fig. 27-5. One type of tube bender. (*Lisle Corporation*)

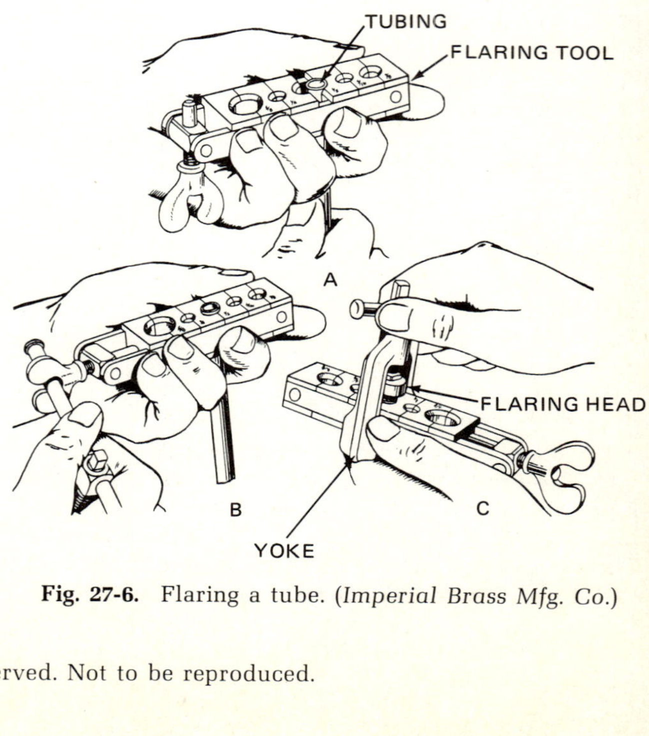

Fig. 27-6. Flaring a tube. (*Imperial Brass Mfg. Co.*)

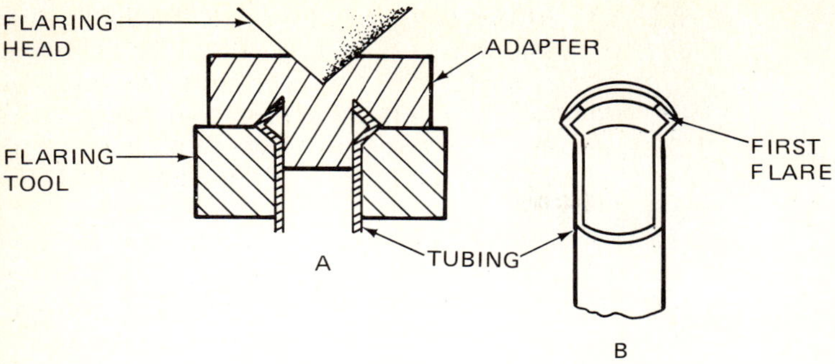

Fig. 27-7. Using the adapter to make the first flare in the double-flare procedure. (*Ford Motor Company*)

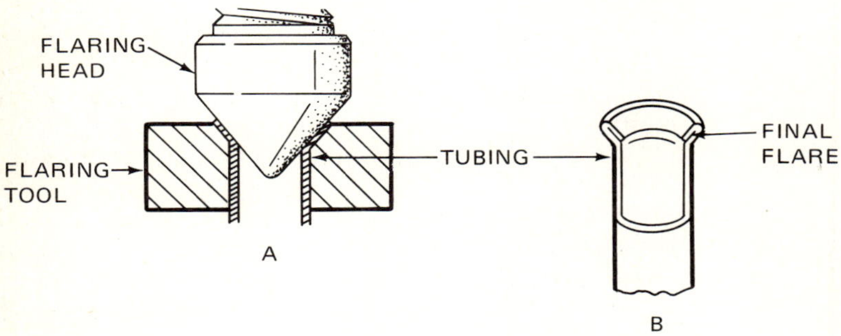

Fig. 27-8. Making the final flare in the double-flare procedure. (*Ford Motor Company*)

_____ **17** Place the flaring head directly in the tubing, and tighten the handle (Fig. 27-6). As the flaring head goes down into the tubing, the head will bend down the first flare, making a tight double flare (Fig. 27-8). Remove the flaring head and yoke, and inspect the flare (Fig. 27-8). If it is okay and without cracks, be sure all fittings or nuts are in place. Then, flare the other end of the tube.

_____ **18** Fasten a piece of copper tubing in a vise, being careful not to bend or crush the tubing. Then mark a spot on the tubing where you want to cut it.

_____ **19** Place the cutting wheel of a tubing cutter on the mark, and tighten the adjusting screw (Fig. 27-2). Be careful not to overtighten the cutting wheel, as it could crush or split the tubing.

_____ **20** Rotate the cutter around the tubing (Fig. 27-3). Tighten the adjusting screw just enough to keep the cutting wheel in contact with the tubing. When the end of the tubing is cut off, use the reamer on the tube cutter to remove the burr from inside the tube (Fig. 27-4). File the end square and smooth. Be sure to remove all filings and chips from inside the tubing.

_____ **21** Loosen the vise, and remove the tubing. Try to bend it gently with your hands, while you watch what happens to the tube. Stop bending as soon as the tubing splits or begins to kink or flatten. Examine the tubing closely, to see why you should not bend tubing without a bending tool.

_____ **22** Now, place the tubing in a tube bender (Fig. 27-5), and make at least two bends (Fig. 27-1). Note how the tubing cutter supports the tubing while it is being bent. This keeps the tube from kinking, splitting, or collapsing.

_____ **23** Place a flare nut on one end of the tube. Then, using a flaring tool, make a double flare on the end (Figs. 27-6 to 27-8). Inspect your work, and compare it with Fig. 27-8. If your first flare looks good, flare the other end of the tubing.

_____ **24** Compare a piece of steel tubing to the copper tubing you have been working with. Notice the difference in weight and feel.

_____ **25** Repeat Steps 18 through 23 with the steel tubing.

Multiple-Choice Test — Job 27

1 Copper tubing should not be used for brake lines because it **A** cracks easily **B** is too hard **C** corrodes easily **D** is difficult to flare **1** _____

2 On the automobile, all tubing ends should be **A** single-flared **B** threaded **C** soldered **D** double-flared **2** _____

3 After tubing is cut, the inside must be **A** reamed **B** filed **C** sandpapered **D** washed **3** _____

4 Forming a double flare is an operation that takes **A** four steps **B** three steps **C** two steps **D** one step **4** _____

5 The problem with a single flare is that it **A** cracks easily **B** is too hard **C** corrodes easily **D** is difficult to flare **5** _____

Written Checkup — Job 27

Write down the differences that you noticed between working with the copper tubing and working with the steel tubing.

NAME DATE CLASS

Objective: To learn how to use air tools and impact wrenches.

Equipment: Practice engine.

Tools: Shop air tools.

Job 28

AIR TOOLS AND IMPACT WRENCHES

Safety: *Compressed-air tools are safe and reliable when properly and sensibly used. However, compressed air can be dangerous, even deadly, when it is misused. Never look into the air outlet or nozzle of any compressed-air tool or device such as a blowgun. Never point discharging air at any part of your body or at anyone else. When you do, you change a useful labor-saving tool into a dangerous weapon that can cause permanent injury and even death.*

PROCEDURE

_____ **1** Tools and equipment operated by compressed air are *pneumatic* devices. "Pneumatic" means "of or related to air." We use the common terms "air tools" and "air equipment" when talking about air-powered shop tools.

_____ **2** In this jobsheet, we describe the tools that are powered by compressed air. But first, let us see how to get air pressure in the shop. Air pressure is produced by an air compressor (Fig. 28-1). The air-compressor unit includes an electric motor that drives an air pump. This pump starts running automatically when the pressure falls below a preset value. The pump sends air into the compressor air tank for storage. That way, there is always a reserve of compressed air ready to serve the air tools and equipment in the shop. A schematic of a complete shop air-supply system is shown in Fig. 28-1.

_____ **3** Air tools give either rotary or reciprocating (back-and-forth) motion. Rotary motion is needed to turn nuts or bolts. For example, Fig. 28-2 shows an air impact wrench being used to loosen a bolt.

_____ **4** Reciprocating motion is needed to drive a cutting or hammering tool. Figure 28-3 shows an air hammer being used to cut a nut that has rusted onto a stud. The air hammer repeatedly and rapidly drives a chisel against the nut. Air hammers can be used with a variety of tools—cutters, chisels, and punches—to do many jobs.

_____ **5** Impact wrenches are used to run nuts and bolts in and out rapidly. The direction of rotation can be changed with a reversing control on the impact wrench.

_____ **6** Most shop impact wrenches have a

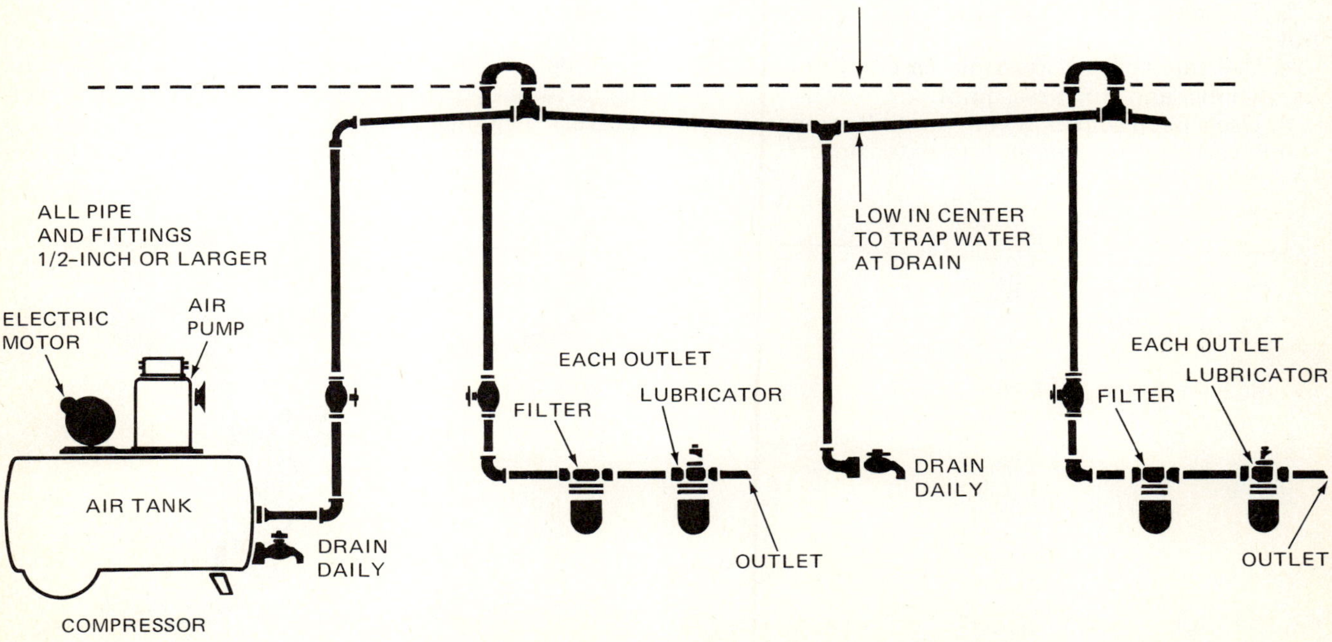

Fig. 28-1. Schematic diagram of a shop compressed-air system. (*Chicago Pneumatic Tool Company*)

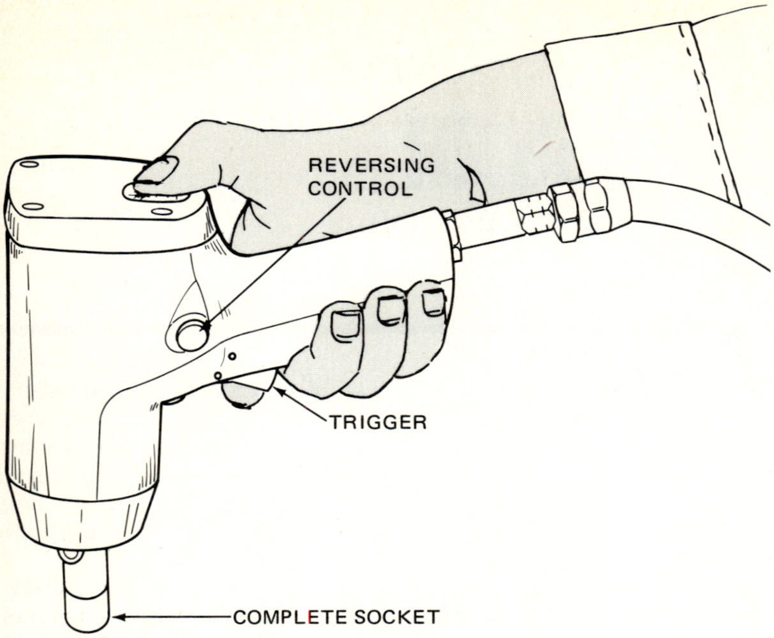

Fig. 28-2. Using an impact wrench.

standard $\frac{1}{2}$-in [12.7-mm] socket drive. However, only special impact sockets should be used with the impact wrench. Standard sockets crack and round out when used on an impact wrench. Impact sockets are frequently black in color.

_____ **7** Some impact wrenches are powered by an electric motor (Fig. 28-4). The electric impact wrench is usually larger and bulkier than a comparable air impact wrench. In addition, you must be careful not to cut or damage the electric cord while working with an electric impact wrench.

_____ **8** Here are some recommendations for using an impact wrench:

a. Always use impact sockets on an impact wrench.

b. Use only the correct size socket for the bolt or nut.

c. Use the simplest possible assembly of sockets, extensions, and universal joint.

d. Use a deep socket, where possible, rather than a standard-length socket and an extension.

e. Hold the wrench so the socket fits squarely on the bolt or nut. Apply a slight forward pressure to hold the socket in place.

f. Once a nut or bolt tightens, never impact it beyond an additional one-half turn of the socket. Continued impacting might strip the threads or break the bolt.

g. Never use an impact wrench beyond its rated capacity. If a bolt or nut remains tight after impacting for 5 seconds, use a larger wrench.

h. Soak large rusty nuts and bolts with penetrating oil before impacting them.

_____ **9** Figure 28-5 shows an air drill with hose and coupling. This drill has a $\frac{1}{4}$-in [6.35-mm] chuck capacity, and a free speed of 2,800 rpm (revolutions per minute). One advantage to the air drill is that it is very light for the amount of power it develops. The air drill shown in Fig. 28-5 develops $\frac{1}{2}$ hp (horsepower). Another advantage of the air drill is that repeated overloading and stalling do not damage or

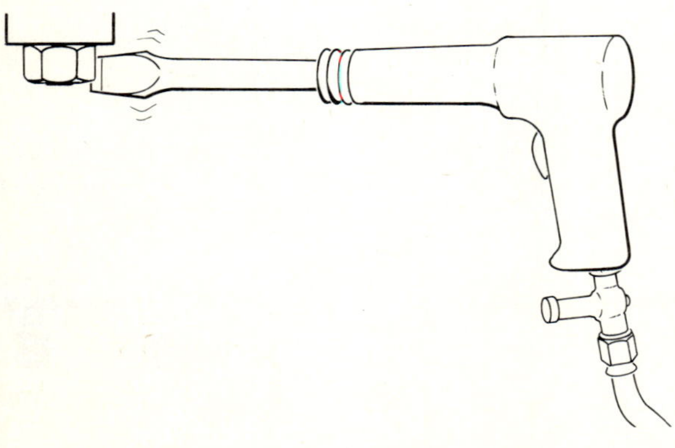
Fig. 28-3. Air hammer in operation, driving a chisel to cut a nut that has frozen onto a stud. (*Chicago Pneumatic Tool Company*)

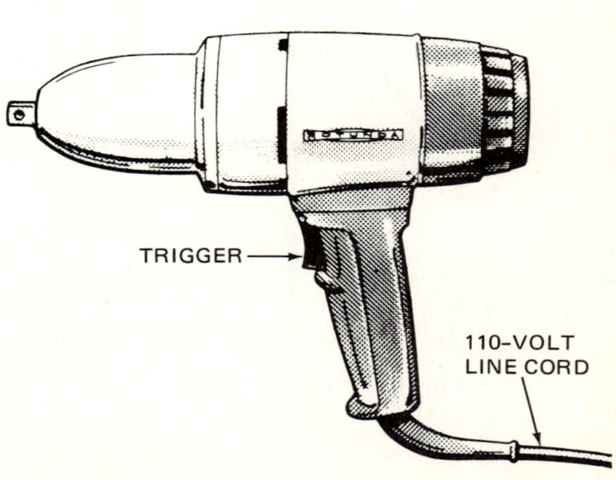

Fig. 28-4. Electric impact wrench. (*Ford Motor Company*)

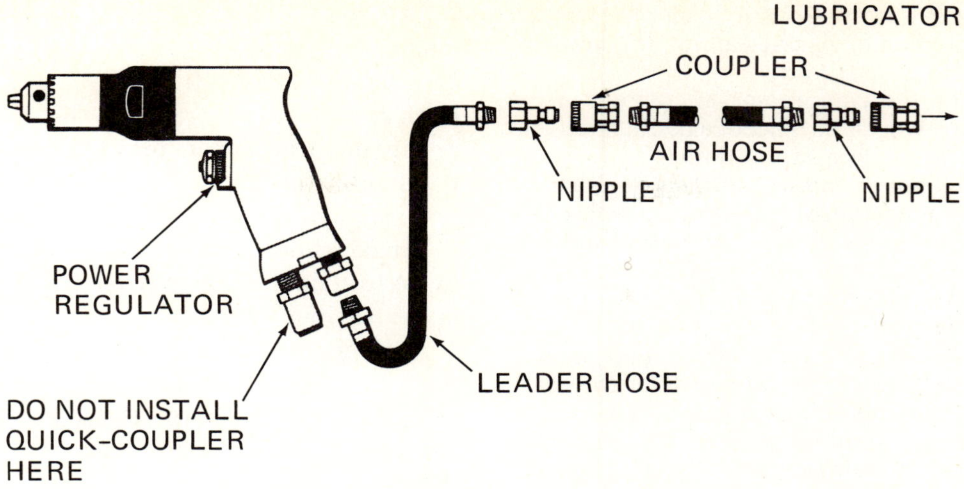

Fig. 28-5. Air drill with hose and couplings. (*Chicago Pneumatic Tool Company*)

overheat it. This is not true for an electric drill.

_____ **10** Air ratchets (Fig. 28-6) are gaining popularity in the shop. They are made with a $\frac{3}{8}$-in [9.53-mm] drive, and can use the sockets and attachments from a standard socket set. Special impact sockets are not required.

_____ **11** For proper performance, any air-powered tool requires a regulated volume of clean, dry air. Most air tools are rated for use with an air pressure of about 100 psi (pounds per square inch) [7.03 kg/cm^2 (kilograms per square centimeter)]. The rated pressure is the pressure at the tool, with the tool running. The tool air pressure is measured by attaching a pressure gauge as close as possible to the tool, and pressing the trigger. The air pressure that registers on the pressure gauge is the pressure at the tool, not the pressure of the shop air compressor.

_____ **12** If the air pressure is less than recommended, the efficiency of an air-powered tool is reduced. However, excessively high air pressure forces the tool to work beyond its rated capacity. This shortens the life of the tool and may cause it to break.

_____ **13** Every day, before using an air tool or impact wrench, apply three or four squirts of flushing oil into the tool air inlet (Fig. 28-7). Then connect and operate the tool. This flushes any gum, dirt, or moisture out of the air motor and lubricates the moving parts. Any deposits in the wrench are flushed out

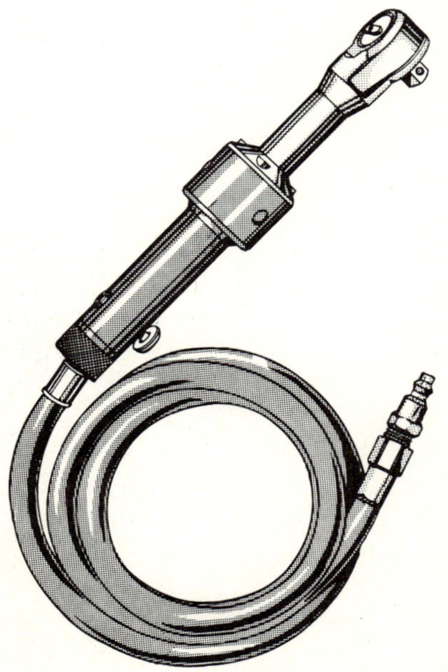

Fig. 28-6. Air ratchet. (*Snap-on Tools Corporation*)

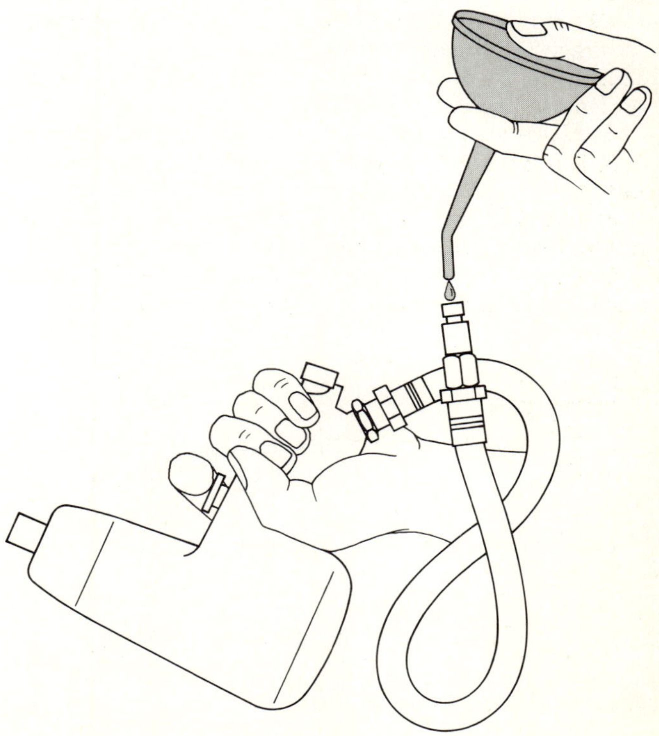

Fig. 28-7. Squirting a few drops of flushing oil into an impact-wrench air line. (*Chicago Pneumatic Tool Company*)

through the air exhaust. Do not flush the wrench around an open flame. Always point the tool air exhaust away from your body.

_____ **14** Locate the compressed-air outlets in the shop and in your work area. Practice connecting and disconnecting an air hose at the outlet.

_____ **15** Check out an impact-wrench set. Identify the trigger, reversing control, and air or power regulator. Squirt a few drops of gum-solvent oil into the air inlet (Fig. 28-7). Connect the hose to the impact wrench, and press the trigger. Then reverse the direction of rotation with the reversing control. Change the position of the power regulator to note how it changes the operation of the impact wrench.

_____ **16** Examine the impact sockets. Compare an impact socket with a standard socket. Note that most impact sockets are 6-point sockets (Job 17). The impact socket is often heavier and slightly thicker than a comparable standard socket.

_____ **17** Practice using the impact wrench on an engine or some other assembly with several sizes of nuts and bolts. Run the bolts in and out. Remember, the impact wrench is a time saver. But for accurate tightening of nuts and bolts, a torque wrench is still required (Job 18).

Multiple-Choice Test — Job 28

1 In the shop, the source of compressed air is **A** the shop air compressor **B** bottles of compressed air **1** _____

2 Air tools provide either **A** rotary or circular motion **B** rotary or reciprocating motion **C** rotary or rotating motion **D** reciprocating or back-and-forth motion **2** _____

3 An impact wrench provides **A** rotary motion **B** reciprocating motion **C** back-and-forth motion **3** _____

4 An advantage of an air drill is that it is undamaged by **A** use under water **B** lack of service **C** water in the air line **D** stalling **4** _____

5 Most air tools require an air pressure of about **A** 60 psi [4.22 kg/cm^2] **B** 100 psi [7.03 kg/cm^2] **C** 150 psi [10.55 kg/cm^2] **D** 250 psi [17.58 kg/cm^2] **5** _____

Written Checkup — Job 28

Inspect the shop air compressor. Note how to turn it on and off, and how to drain it. Locate the compressor pressure gauge, and note the reading. Then write down in your notebook the compressor startup and servicing procedure.

CAUTION: The compressor is automatically controlled and may start running at any time. Keep your hands and clothing away from the motor and drive belts.

NAME DATE CLASS

Objective: To learn how to use electric drills.

Equipment: Various samples of round and flat metal stock through which holes can be drilled.

Tools: Electric drill, set of twist drills, drill gauge, drill stand or drill press, vise, V block, hammer, center punch, scriber, C clamp.

Job 29

ELECTRIC DRILLS AND TWIST DRILLS

Safety: *When using a portable electric drill, be sure to keep a firm grip on it. This will prevent it from jerking out of your hands if the drill jams. If it did jerk loose, the twist drill might snap. The drill would jump against you, giving you a severe cut or bruise. With any drill, make sure that the twist drill is tight in the chuck. Observe the other precautions against drill breakage, as outlined below. Make sure the electric cord to the drill is in good condition. Never kink the cord or allow anyone to step on it or run any machine over it. Any of these might damage the cord insulation so that someone would be shocked when handling it. Put the drill away or on the bench when you are not using it. Never allow it to lie on the floor where someone could step on it or trip over it. Never drag it around by the cord.*

CAUTION: Do not overload a drill so much that it stalls. If a drill does stall, work the twist drill loose carefully, to avoid breaking it off. Never click the switch on and off in an attempt to start a stalled drill. This may burn out the switch or the motor.

PROCEDURE

_____ **1** Drills are tools for making holes. The type of material into which the hole is to be drilled determines the type of drill to be used. The typical drill, the *twist drill*, is a round bar with helical grooves (Fig. 29-1). Note that the drill has two cutting edges. Chips pass up the helical grooves and away from the working area.

_____ **2** Twist drills are often used with hand-operated drills (Fig. 29-2). However, most drills for the shop are powered by electric motors (Fig. 29-3). An electric drill contains a set of reduction gears that turn the twist drill at a relatively slow speed.

_____ **3** For more accurate drilling and for larger holes, a drill press is used (Fig. 29-4). This has a heavy upright that supports the drill motor and drill. It also has a table on which the work can be placed and attached, if necessary. A handle is used to bring the drill down to the work. The handle turns a gear that moves a rack down.

_____ **4** A center punch should always be used to mark the location of a hole to be drilled in a metal piece. The center-punch mark locates the hole accurately and keeps the drill from wandering (Fig. 29-5).

_____ **5** Here are some cautions to observe when using electric drills and other power tools:

a. Read and understand the instructions before you operate the power tool.

b. Have your instructor check you out and give you permission to operate the tool.

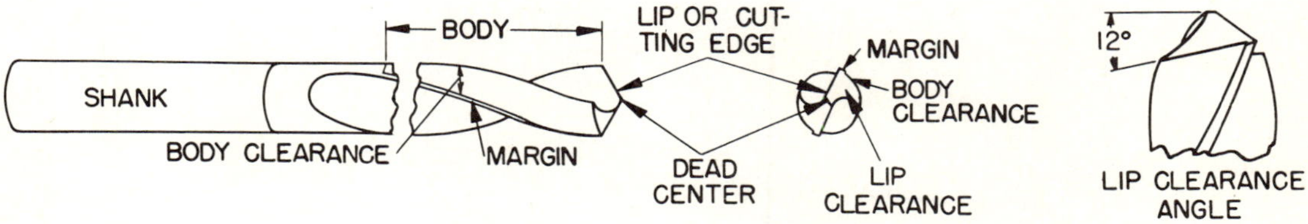

Fig. 29-1. Parts of a twist drill.

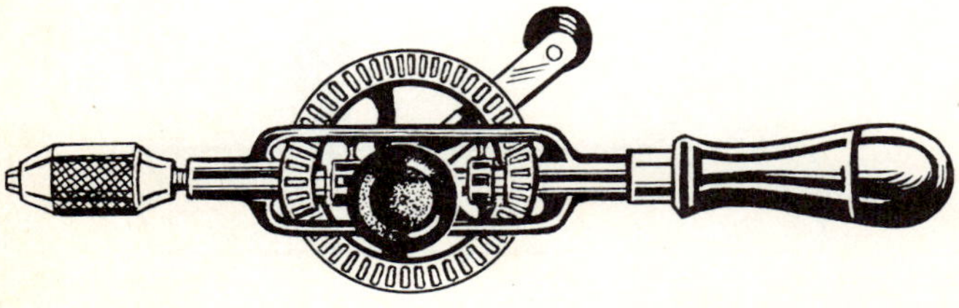

Fig. 29-2. Hand drill.

c. Keep your hands and clothing away from moving parts.

d. Make sure that the work is securely held so that nothing can come loose when the tool is turned on.

e. Be ready to turn the tool off quickly if something seems wrong. Never take a chance; be sure everything is in good order before proceeding.

f. Do not drag portable power tools, such as drills (Fig. 29-3), around by the electric cord.

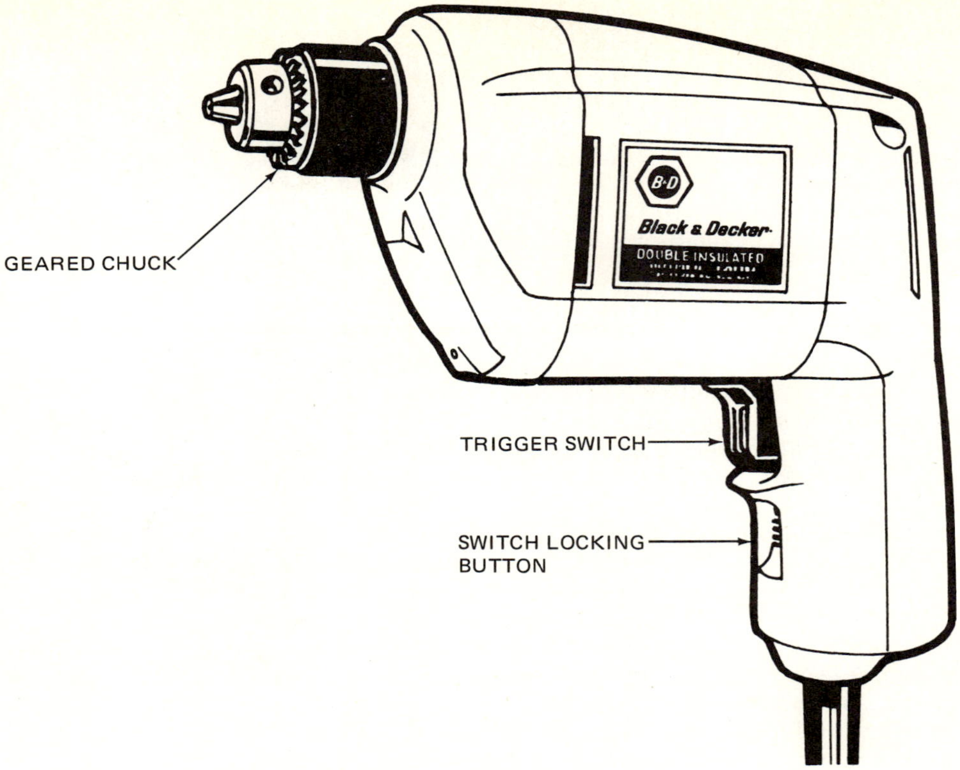

Fig. 29-3. Electric drill. (*The Black & Decker Manufacturing Company*)

g. Always make sure that portable power tools are grounded (Fig. 29-6). Make sure the cord is not in someone's way.

h. After you use a portable tool, clean it and return it to the bench or tool room where it belongs.

i. Never attempt to adjust or oil moving machinery unless the instructions specifically state that this should be done.

_____ **6** Examine a twist drill, and note its various parts (Fig. 29-1). The twist drill can be driven either by a hand drill or by an electric drill. The drilling fundamentals are the same for both hand-operated and electric drills.

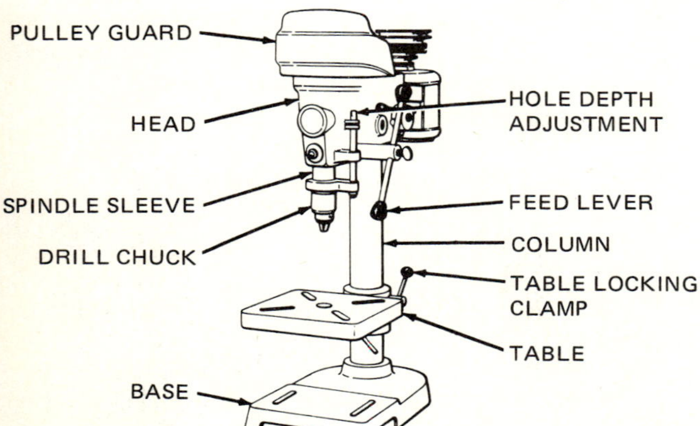

Fig. 29-4. Drill press. (*Ford Motor Company*)

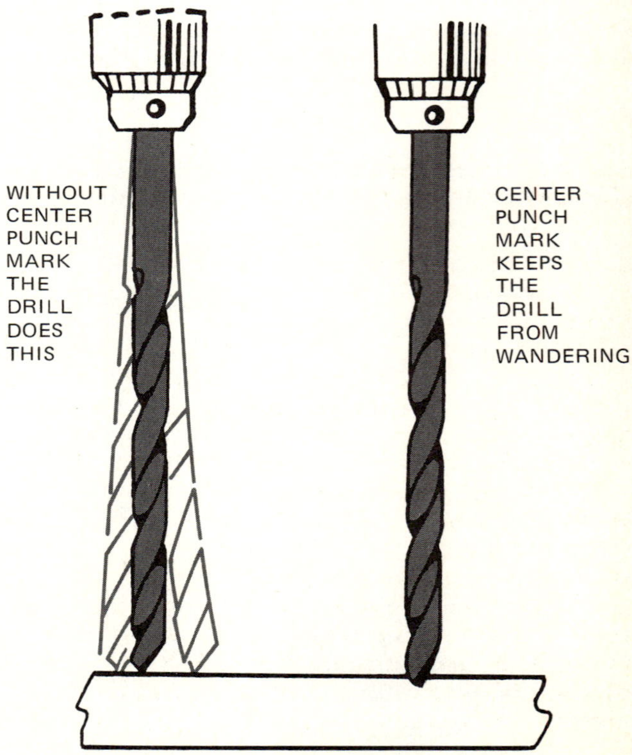

Fig. 29-5. How center-punching a hole location keeps the drill from wandering. (*General Motors Corporation*)

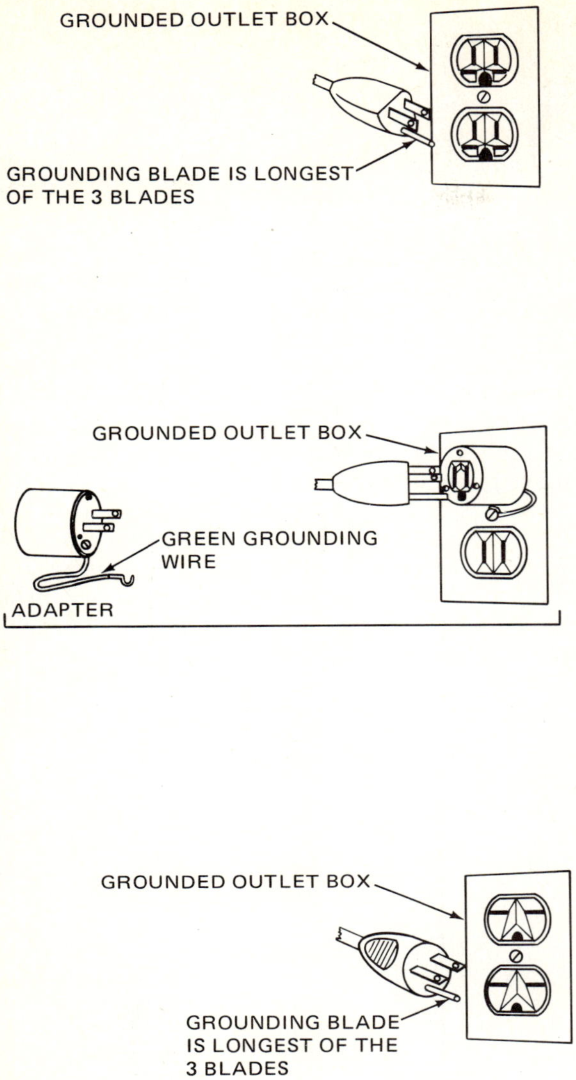

Fig. 29-6. Various methods of grounding electric tools. (*The Black & Decker Manufacturing Company*)

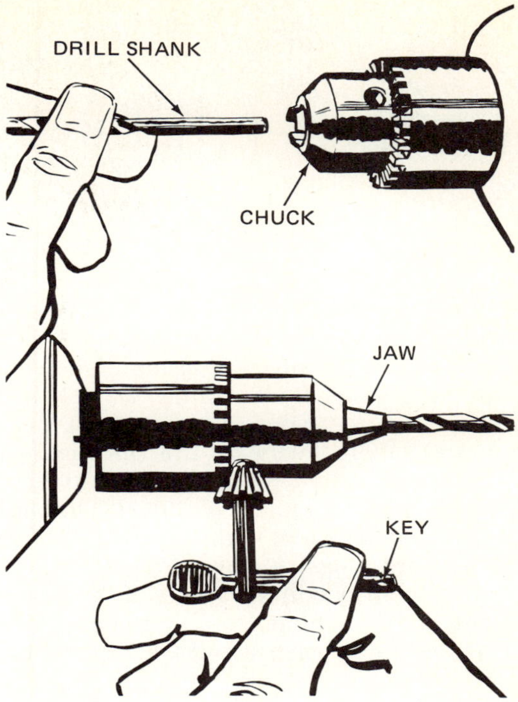

Fig. 29-7. Installing a twist drill in a drill chuck.

_____ **7** Examine an electric drill (Fig. 29-3). Note how it is held, started, and stopped. If possible, your instructor should demonstrate how to hold and use the electric drill.

_____ **8** Put a twist drill into the chuck of an electric drill, and tighten the chuck firmly (Fig. 29-7). Make sure the drill is inserted into the chuck as far as it will go.

_____ **9** Select a piece of metal stock. Use a scriber to scratch intersecting lines on the metal where holes are to be drilled. Then center-punch the stock at the intersections (Job 19).

_____ **10** Clamp the metal stock in a vise. Connect the drill, and turn it on (Fig. 29-8). Apply the point of the drill to one center-punch mark. Keep the drill square with the surface of the stock. Apply a steady pressure to the drill.

_____ **11** Before the drill point has cut all the way into the metal, lift the drill. Note whether the drill mark is centered on the intersecting lines. If it is not, use a round-nose chisel to make a cut on the side toward which the drill should shift (Fig. 29-9). One or more chisel cuts will draw the drill to that side so that it will center correctly.

_____ **12** Proceed with the drilling operation. Apply steady pressure, and keep the drill square with the surface of the stock.

_____ **13** As the drill begins to break through the metal, ease up on the pressure. Take a firmer hold on the drill. Sometimes, the drill tends to jam as it goes through, and you should be on guard for this.

_____ **14** Do not try to drill too fast or too slowly. This overheats the drill and destroys its temper. The drill becomes soft and dull and does not cut properly. A few drops of oil or cutting compound will help the cutting action. Put the oil on the side of the drill where it will work down into the hole being drilled.

_____ **15** Mount the electric drill in a drill stand, or use the drill press. Practice drilling holes in flat and

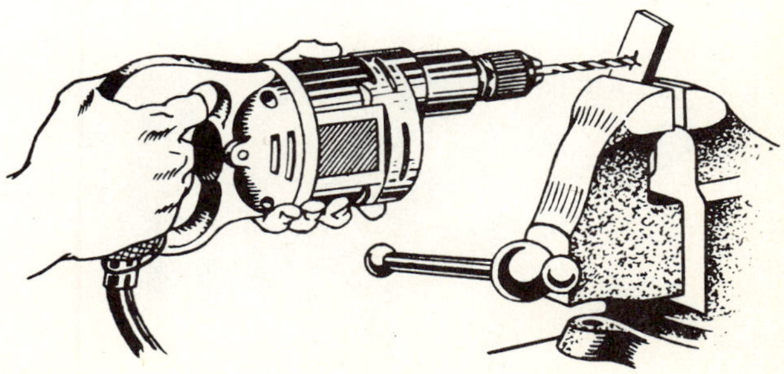

Fig. 29-8. Drilling a small piece of metal held in a vise.

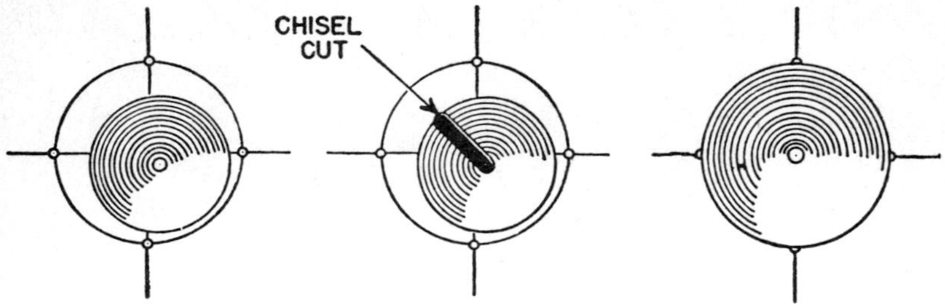

Fig. 29-9. If the drill starts off center, it may be drawn back to center with a chisel mark made on the side toward which the drill should be drawn.

round stock. Be sure that the press table is positioned so that the drill can pass through a hole in the table as it completes the drilling job. Use a V block to keep round stock from rolling while it is being drilled (Fig. 29-10). A block of wood with a V-shaped notch serves satisfactorily for this purpose.

_____ **16** To drill thin metal, place the work on a block of wood for support (Fig. 29-11). Then use a C clamp to secure the metal while you drill it. This will reduce drill breakage.

_____ **17** Clamp a bolt vertically in a vise. Use a center punch and hammer to punch-mark the exact center of the bolt. Select a twist drill smaller than the root diameter of the bolt threads. Put the twist drill into a hand drill, and drill a hole vertically down through the bolt. Work carefully to keep the drill centered in the bolt. Do not allow the drill to angle in the bolt.

_____ **18** Drill the bolt about 1½ to 2 in [38.1 to 50.8 mm] deep. Then take out the drill. Saw off the drilled section of the bolt, with a hacksaw (Job 22). If the bottom of the hole is also centered in the bolt, you have done a good drilling job. It is important to be able to do this job correctly, to remove broken studs (Job 26). Broken studs must be drilled before they can be removed. Drilling a broken stud in a

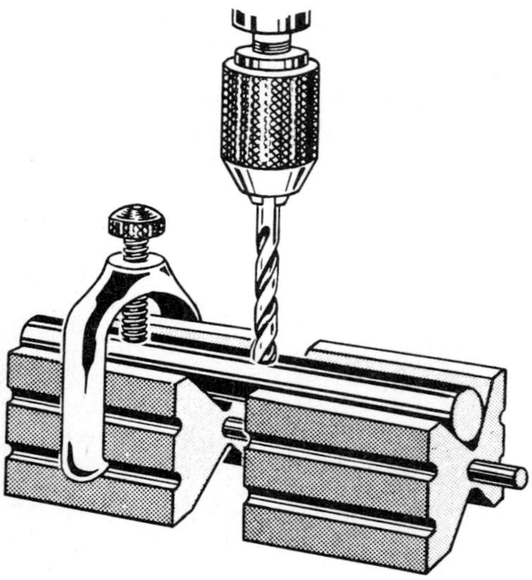

Fig. 29-10. Drilling a rod held in V blocks.

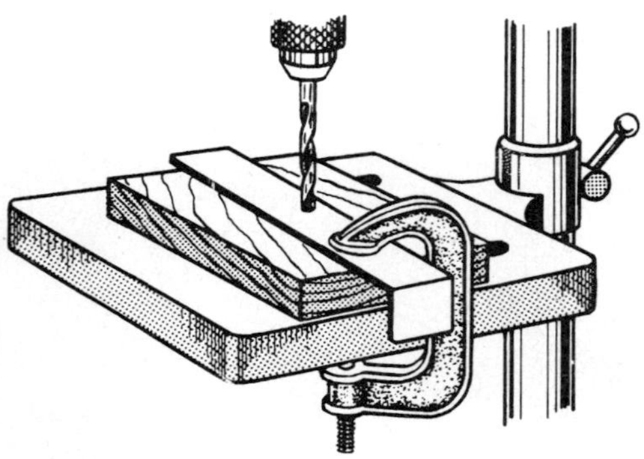

Fig. 29-11. To drill holes in thin sheet metal, clamp the sheet metal in place over a piece of wood.

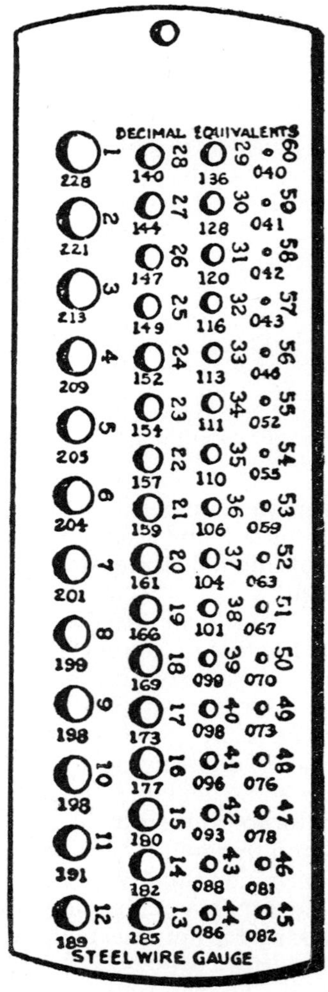

Fig. 29-12. Drill gauge. The number and size of the drill are shown beside the hole in which the drill fits snugly.

cylinder block at an angle may damage the cylinder block, because the drill might enter the block.

_____ **19** Twist drills should never be kept loose in a box or a drawer. The cutting edges can become banged up and dulled. Instead, they should be kept in a rack or a divided box. Drill sharpening is detailed in Job 30.

_____ **20** The electric drill should be put away after each use. It should be kept clean and lubricated, according to the manufacturer's instructions.

_____ **21** Use a drill gauge (Fig. 29-12) to check the sizes of several twist drills. To do this, find the gauge hole into which the drill fits snugly. Then note the drill size, as marked at the side of the hole. In the space below, record the drill sizes that you check.

MEASURING DRILLS WITH A DRILL GAUGE	
Drill 1	_____
Drill 2	_____
Drill 3	_____
Drill 4	_____
Drill 5	_____

True-or-False Test — Job 29

1 The shank of a twist drill contains the cutting edges of the drill. **1** _____

2 The cord of an electric drill must not be kinked or run over with a machine. **2** _____

3 The twist drill sometimes jams as it completes a cut. **3** _____

4 A center-punch mark helps to center the drill and prevent it from wandering. **4** _____

5 Excessively fast drilling may destroy the drill temper. **5** _____

6 A drill gauge is used to gauge the drill-hole depth. **6** _____

7 If a drill jams, the switch should be clicked off and on several times to free it. **7** _____

8 A drill chuck has jaws that hold the twist drill when tightened. **8** _____

9 If the drill mark is not centered on the center lines, you should make a cut with a chisel on the side away from which the drill must shift. **9** _____

10 A V block should be used to support round stock when it is being drilled. **10** _____

Written Checkup — Job 29

1 Write down the procedure for marking and drilling a hole in a piece of stock.

2 Write down the procedure for drilling a hole in a piece of round stock.

NAME DATE CLASS

Objective: To learn how to use the bench grinder.

Equipment: Sections of round and flat metal stock.

Tools: Grinder, grinding-wheel dresser, screwdrivers, punches, chisels, drills, drill-grinding fixture.

Job 30

BENCH GRINDERS

Safety: *The grinding wheel is dangerous if improperly used. Always observe the following precautions: Do not hammer on the wheel. Make sure that the tool rest is tight and in proper position before starting the grinder. Never attempt to make tool-rest adjustments with the grinder in operation. Keep a firm grip on the tool or part being ground. Do not touch the rotating wheel with your hand. Always wear safety goggles when using a grinding wheel, even though the wheel is equipped with an eyeshield. Failure to wear goggles is very risky, since flying particles could easily cost you the sight of an eye.*

PROCEDURE

Part One THE BENCH GRINDER

_____ **1** A bench grinder (Fig. 30-1) has one or more grinding wheels made of abrasive material that is bonded together. When the wheel is rotated by the electric motor, objects held against the wheel are ground down. Thus, the grinder can be used to shape and sharpen tools.

_____ **2** Many grinders have a combination of one grinding wheel and one wire wheel (Fig. 30-2). The wire wheel has hundreds of wires fastened to a central core. When the wire wheel revolves, objects held against it are cleaned and polished.

_____ **3** The cutting element of the grinder is the grinding wheel. Grinding wheels (Fig. 30-2) are made of abrasive material. They are made in many sizes and with many different kinds and grades of abrasive. For coarse work, the abrasive particles are relatively large; they cut the material very rapidly. An example might be the grinding wheels used to take the rough edges off metal castings. For fine work, such as sharpening small tools, a relatively fine grinding wheel is used.

_____ **4** Great care must be taken when the grinding wheel is used. The wheel is driven by an electric motor and revolves at fairly high speed. The wheel could be broken by heavy blows, heavy pressure, or excessive tightening of the spindle nut. During the grinding of many metals, showers of hot sparks are thrown off by the wheel. These sparks can set fire to your clothing, or burn you. You must be alert while using a bench grinder.

_____ **5** Goggles should always be worn when the grinding wheel is used, and the safety shield should be in place (Fig. 30-1).

_____ **6** When using the wheel, do not allow the metal being ground to get overheated. This is particularly important when grinding a tool such as a screwdriver, chisel, or drill. Overheating ruins the temper of the steel and makes the tool so soft that

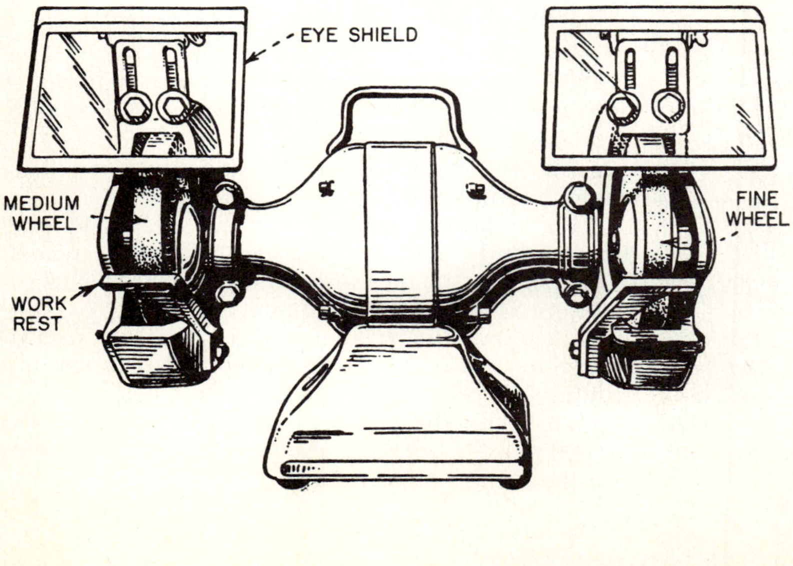

Fig. 30-1. Bench grinder.

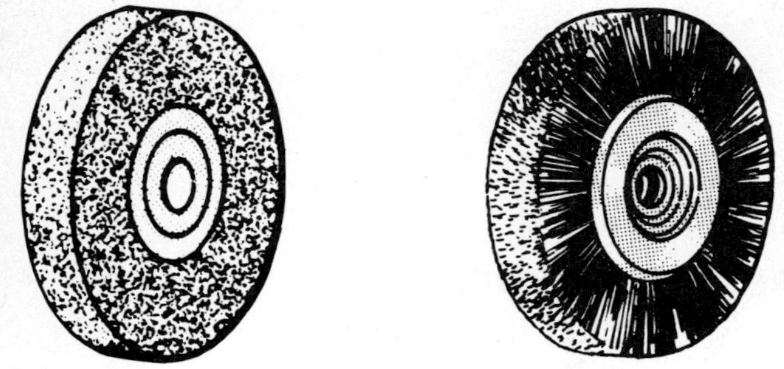

Fig. 30-2. Grinding wheel and wire wheel for a bench grinder. (*Ford Motor Company*)

the edge does not last. To prevent overheating, dip the tool in water repeatedly during the grinding operation.

CAUTION: Do not use the grinder, or any other powered tool, until your instructor has explained how it operates and has told you to go ahead. These tools are easy and safe to use if they are used properly. But they can cause injuries if they are improperly used.

Part Two THE GRINDING WHEEL

_____ **1** Examine the grinding wheel before grinding screwdrivers, punches, chisels, and drills.
_____ **2** Many grinders have two wheels (Fig. 30-1)—a coarse-grained wheel for rough work, and a fine-grained wheel for fine and finishing work. Other grinders have a grinding wheel and a wire brush (Fig. 30-2). If the grinder has two wheels, examine them and note the difference in the grains.
_____ **3** If the grinding surface of the wheel is rough or grooved, it should be dressed. Your instructor will show you how to use the wheel dresser (Fig. 30-3).

Fig. 30-3. Dressing the grinding wheel with a wheel dresser.

NOTE: When using a grinding wheel, always move the piece being ground back and forth on the grinding surface. If the piece is held in one spot, a groove may wear in the wheel.

Part Three USING THE BENCH GRINDER

_____ **1** Select a piece of stock, several inches long, to practice on. Adjust the tool rest so that it is at right angles to the grinding surface of the wheel. Tighten the rest.
_____ **2** Put the eyeshield in position, and put your goggles on.
_____ **3** Start the grinder. Lay the stock on the rest so that the end is almost touching the wheel. Hold the stock firmly, and push it into the wheel (Fig. 30-4). Note how the end is ground off. Rotate the stock to grind the end flat.
_____ **4** Now lay the stock on the rest so that a chamfer is ground on the end. Rotate the stock slowly, to grind its end to a smaller diameter. When you finish, your stock should look something like the end of the chisel in Fig. 30-5.
_____ **5** Notice that the stock becomes hot as it is ground. Dip it repeatedly in water to keep it from overheating. Overheating a cutting tool during grinding ruins its temper, and it becomes soft and *useless*.
_____ **6** Select another piece of stock. Grind a point on one end. Try to make it look like a center punch (Fig. 30-7).

Fig. 30-4. Grinding the end of a piece of bar stock.

Part Four GRINDING A SCREWDRIVER

_____ **1** Adjust the rest to hold a screwdriver at right angles to the grinding surface of the wheel (Fig. 30-5). Tighten the rest.
_____ **2** Put your goggles on, and lower the eyeshield into position.
_____ **3** Start the grinder. Lay the screwdriver on the rest, and feed the tip squarely into the wheel. Grind off the tip no more than is necessary to remove nicks and square the end.
_____ **4** Stop the grinder. Adjust the tool rest to hold the side of the screwdriver tip against the grinding surface (Fig. 30-5).
_____ **5** Start the grinder. Lay the screwdriver on the rest and the wheel as shown. Grind both sides to give the tip the proper shape.

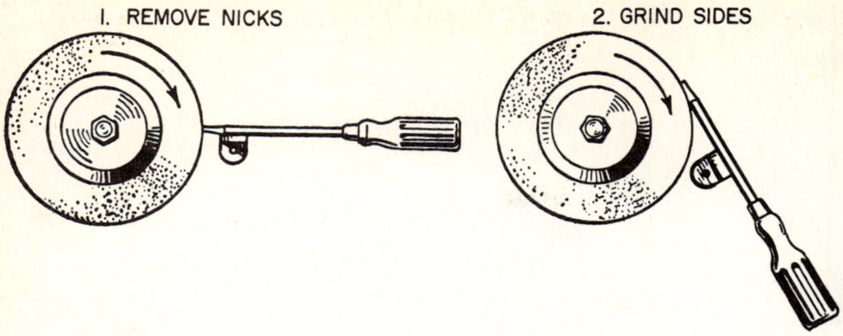

Fig. 30-5. Procedure for grinding a screwdriver.

_____ **6** Dip the screwdriver tip into water repeatedly during the grinding, to prevent it from overheating and losing its temper.

_____ **7** After you have become experienced in using a grinding wheel, you will find it possible to hold and guide the screwdriver in both hands. You will not need to use the rest, except, perhaps, to support your left hand.

Part Five GRINDING PUNCHES

_____ **1** A blunt-end punch, such as a pin or taper punch, must have a perfectly flat end, at right angles to its centerline. To grind the end of such a punch, adjust the rest so that it is at right angles to the wheel grinding surface. Then, with your goggles on and the eyeshield in place, start the grinder. Feed the punch into the grinder at right angles to the grinding surface. Rotate the punch to get a flat end. Do not take off any more metal than is necessary. Repeatedly dip the punch into water to avoid overheating it.

_____ **2** Pointed punches (center and prick punches) should have their points sharpened to the proper angles (Fig. 30-6). Adjust the rest so that the punch end meets the wheel at the correct angle. Then put on your goggles, adjust the guard, and turn on the grinder. Lay the punch on the rest and the wheel. Rotate the punch, and dip it into water frequently to avoid burning the point.

_____ **3** The heads of punches and chisels may mushroom after considerable use (Fig. 30-7). In use, bits of the mushroom could break off and fly away at high speed. Such particles are dangerous and could cause a serious eye or hand injury. For this reason, such punches should never be used. The mushroom should be ground off as follows:

_____ **4** Adjust the rest to a horizontal position. Put your goggles on, and lower the eyeshield. Start the grinder. Lay the tool on the guard so that the mushroom is ground off and the original shape of the head is restored (Fig. 30-8). Dip the punch in water repeatedly, to avoid overheating.

_____ **5** After a punch or chisel has been ground, it must be hardened and tempered so that the end and the head will hold their shapes.

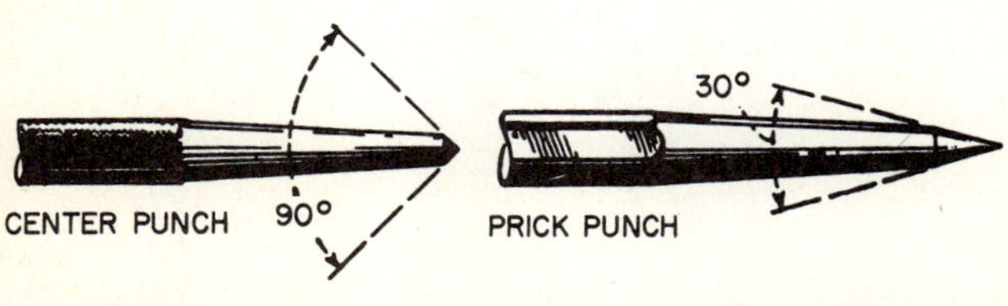

Fig. 30-6. Proper angles to which the points of center and prick punches should be ground.

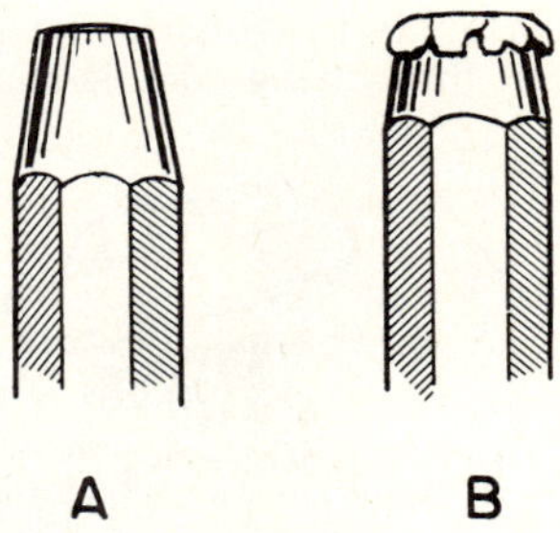

Fig. 30-7. (*A*) Properly dressed head of a punch. (*B*) Mushroomed head of a punch.

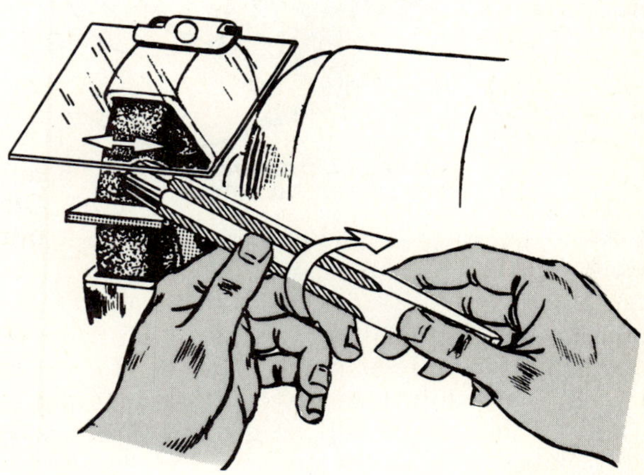

Fig. 30-8. Grinding the mushroom from the head of a chisel.

Part Six GRINDING CHISELS

_____ **1** The cutting angle of a chisel varies somewhat with the material to be cut, but in general it is 70° (Fig. 30-9). Note also that the cutting edge is slightly curved.

_____ **2** To grind a chisel, first remove any nicks

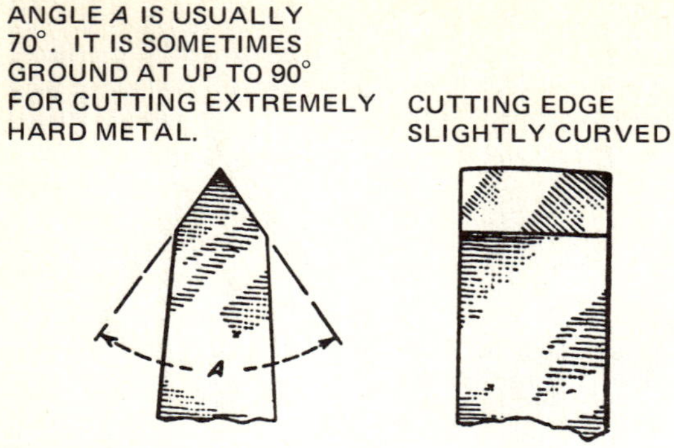

Fig. 30-9. Proper angle to which a flat chisel should be ground. Note that the cutting edge is slightly curved.

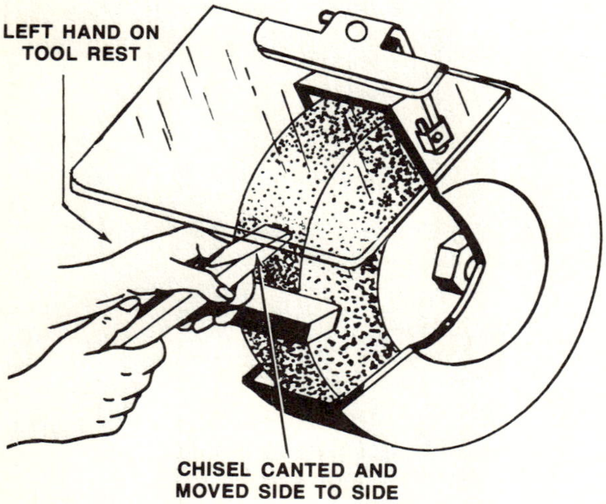

Fig. 30-10. Dressing or grinding a chisel. (*General Motors Corporation*)

from the cutting edge, just as you did on the screwdriver. Remove only as much metal as is necessary.

_____ **3** Then adjust the rest (with grinder stopped) so that the chisel tip will be ground to the proper angle. Put your goggles on, and the eyeshield in place. Lay the chisel on the rest (Fig. 30-10) with the wheel rotating. Grind a new cutting edge. This is done in almost the same way as you ground the sides of the screwdriver tip. However, you have to pivot the chisel on the rest, and swing the end being ground back and forth across the wheel. This will curve the cutting edge slightly (Fig. 30-9).

Turn the chisel over, and repeat on other side. When finished, the bevels on both sides of the cutting edge should be of the same size. Dip the tip into water repeatedly, to prevent overheating.

_____ **4** Remove the mushroom if present (Fig. 30-8). Harden and temper the chisel after grinding.

Part Seven GRINDING DRILLS

_____ **1** A drill-grinding fixture is best for this job. Without one, it is difficult to get the exact angles required.

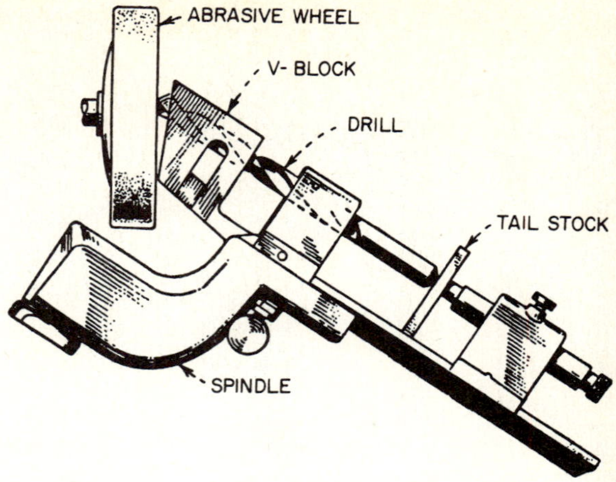

Fig. 30-11. Drill-grinding fixture for holding a drill while it is being ground.

_____ **2** To use the fixture, adjust it to proper cutting-edge and heel angles (Fig. 30-11). Place the drill in the holder, and turn it so that the cutting edge feeds into the wheel. Start the grinder, with your goggles on and the eyeshield in place. Feed the drill into the grinding surface. Swing the holder through the arc to grind one cutting edge of the drill. Then, without changing the holder adjustment, rotate the drill ½ turn. Repeat to sharpen the other cutting edge.

_____ **3** If a fixture is not available, the drill may be ground by hand. But you must be careful to obtain the proper heel and lip angles (Fig. 30-12). The use of a drill gauge (Fig. 30-13) will help you establish the proper angles.

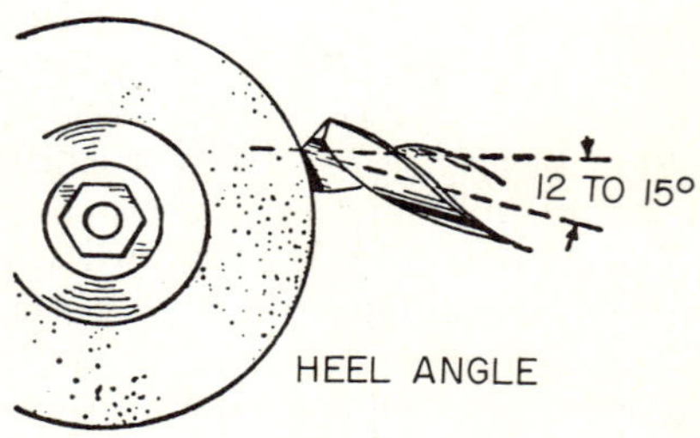

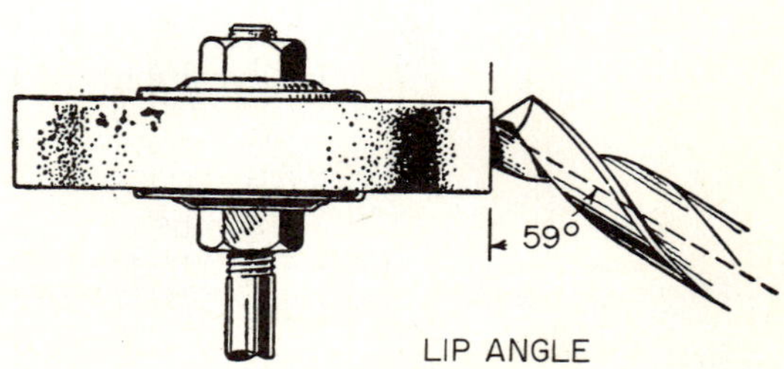

Fig. 30-12. Angle at which the drill should be held on the wheel to obtain the proper heel and lip angles on the end of the drill.

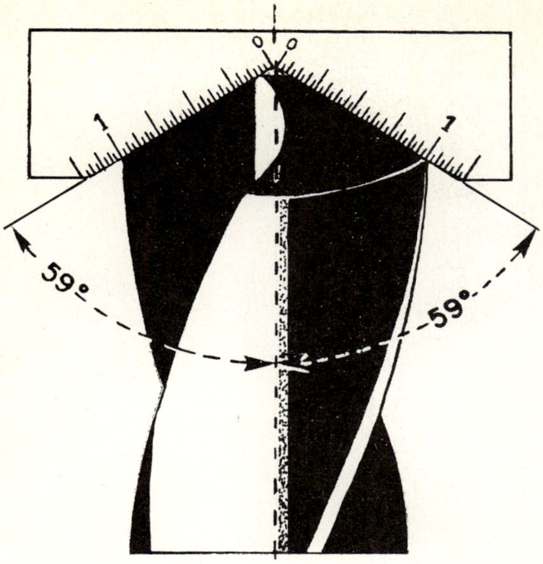

Fig. 30-13. Gauge which can be used to check drill angles.

True-or-False Test — Job 30

1 Anyone using a grinding wheel should wear goggles. 1 ______

2 The tool rest should never be adjusted while the wheel is in motion. 2 ______

3 The purpose of dressing the wheel is to add more abrasive, thereby smoothing the wheel. 3 ______

4 When tools are being ground, they should be repeatedly dipped into water so that they do not overheat. 4 ______

5 The cutting edge of a flat chisel should be slightly curved. 5 ______

6 The heel angle of a drill should be 59°. 6 ______

7 The faces of a screwdriver tip should be at about 90° angles. 7 ______

8 For most work, the twist-drill lip angle should be 59° 8 ______

9 Most grinding wheels rotate at about 10 rpm. 9 ______

10 A rotating grinding wheel should never be touched with the hand. 10 ______

Written Checkup — Job 30

1 Explain why goggles should always be worn when a grinding wheel is being used.

2 Write down the procedure for grinding a screwdriver tip.

3 Write down the procedure for grinding a chisel.

4 List the cautions to be observed when using a grinding wheel.

NAME DATE CLASS

Objective: To learn to use the metric system.

Job 31

USING THE METRIC SYSTEM

PROCEDURE

_____ 1 Suppose you are tuning a car. Some of the specifications are given in the U.S. Customary System in inches, and others are given in the metric system. For the conversions, you check Table 31-1.

_____ 2 The specifications call for the spark plugs to be gapped at 0.035 in. To determine the metric equivalent, you proceed as follows:

$$0.035 \times 25.4 = 0.889 \text{ mm}$$

That is, to change a measurement given in inches to its metric equivalent, multiply the inch measurement by 25.4. This is the number of millimeters in one inch. The answer is the metric equivalent of the inch measurement.

_____ 3 According to the service manual you are using, the contact-point gap should be 0.018 in. Determine the metric equivalent by following the procedure shown in Step 2 above.

$$\underset{\text{inch measurement}}{\underline{\hspace{3cm}}} \times 25.4 = \underline{\hspace{3cm}} \text{ mm}$$

Be careful about placing the decimal point!

_____ 4 With the engine running, you connect a dwell meter and check the reading. The meter reads 30°, which is within the specification of 28 to 32°. Note that there are 360° in a circle, in both the metric system and the U.S. Customary System. Angles, such as ignition-timing, valve-timing, and front-end-alignment angles, are always the same in both systems of measurement.

_____ 5 The decal on the engine air cleaner states that the engine has a displacement of 225 in^3 (cubic inches). You know that in the metric system engine size is measure in liters. One liter is approximately 61 in^3. Therefore, to find the engine size in liters, you use the formula

$$\frac{\text{Engine size in cubic inches}}{\text{Number of cubic inches in one liter}} = \text{engine size in liters}$$

Substituting the size of the engine, you get:

$$\frac{\underline{\hspace{1cm}}}{61} = \underline{\hspace{2cm}} \text{ liters}$$

_____ 6 From the specifications, you note that the engine can run on regular fuel and has a compression ratio of 8.4:1. Numerical ratios are also the same in both systems. An engine compression ratio is not affected by the units used to make engine measurements. The same is true of transmission and rear-axle ratios, which are the same in both systems of measurement.

_____ 7 The horsepower of the engine you are tuning is given as 105 hp at 3,600 rpm. In the metric system, power is measured in kilowatts (kW). To change horsepower to kilowatts, you use the formula

$$\text{hp} \times 0.746 = \text{kW}$$

Find the power of the engine in the metric system by substituting in the formula:

$$\underline{\hspace{2cm}} \times 0.746 = \underline{\hspace{2cm}} \text{ kW}$$

Note that, in the metric system, the units of measure are named in honor of famous scientists. Therefore, the abbreviation of the name Watt (W) is capitalized, while the prefix kilo (k) is not. Hence, kilowatt is abbreviated "kW."

Multiple-Choice Test — Job 31

1 The contact-point gap for an imported car is given as 0.45 mm. In the U.S. Customary System, the points should be gapped at **A** 0.018 in **B** 0.015 in **C** 0.180 in **D** 0.021 in **1** _____

2 The fuel-tank capacity of a car is 21 gallons. The tank holds **A** 7.5 liters **B** 75 cc (cubic centimeters) **C** 75 liters **D** 750 cc **2** _____

3 The bore of an engine is 3.40 in. In the metric system the bore is **A** 86.36 mm **B** 86.36 m (meters) **C** 86.36 cm (centimeters) **D** 86.36 km (kilometers) **3** _____

TABLE 31-1

LENGTH		
1 in (inch)	= 2.54 cm (centimeters)	or 0.0254 m (meter)
1 cm	= 0.39 in	or 0.03281 ft (foot)
1 mm (millimeter)	= 0.039 in	or 0.003281 ft
1 ft	= 30.48 cm	or 0.3048 m
1 mi (mile)	= 1.609 km (kilometers)	or 1,609 m
1 km	= 0.62 mi	or 3,281 ft
VOLUME/CAPACITY		
1 in^3 (cubic inch)	= 16.39 cm^3 (cubic centimeters)	or 0.01 liter
1 cm^3	= 0.061 in^3	or 0.001 liter
1 liter	= 61.02 in^3	or 1.057 qt (quarts)
1 gal (gallon)	= 4 qt	or 3.780 liters
WEIGHT		
1 kg (kilogram)	= 2.2 lb (pounds)	= 35.2 oz (ounces)
1 lb	= 0.454 kg	

Here are the metric measurements, taken from the complete metric-system table, that you will work with most often.

LENGTH		
1 km	= 1,000 m	= 100,000 cm
1 m	= 100 cm	= 1,000 mm (millimeters)
VOLUME/CAPACITY		
1 kl (kiloliter)	= 1,000 liters	= 100,000 cl (centiliters)
1 liter	= 1,000 cm^3	= 1,000 ml (milliliters)
WEIGHT		
1 kg (kilogram)	= 1,000 g (grams)	= 100,000 cg (centigrams)

4 A connecting-rod bearing clearance is given as 0.051 mm. In the U.S. Customary System, the clearance is **A** 0.0002 in **B** 0.2000 in **C** 0.002 in **D** 0.020 in **4** _______

5 In the U.S. Customary System, the camshaft turns at one-half the crankshaft speed. Therefore, in the metric system, the camshaft turns at **A** the crankshaft speed **B** one-fourth the crankshaft speed **C** one-half the crankshaft speed **D** twice the crankshaft speed **5** _______

Written Checkup **Job 31**

On the left side of a sheet of paper from your notebook, write the tuneup specifications for your car in the U.S. Customary System. Then, on the right side of the paper, write the tuneup specifications in the metric system. Use the metric conversion tables in the appendix to find the metric equivalents. File the paper in your notebook for future use.

NAME DATE CLASS

Objective: To learn how to find metric engine measurements.

Job 32

METRIC ENGINE MEASUREMENTS

PROCEDURE

_____ **1** Most of the thermometers we use measure temperature on the Fahrenheit (F) scale. On this scale, water freezes at 32°F and boils at 212°F. With the increasing use of the metric system, temperatures are often given in degrees Celsius (C). The Celsius scale was formerly called the centigrade scale.

_____ **2** A frequent problem for the automobile mechanic is that the thermometer reads in °F and the specification is given in °C. To convert from °F to °C, you use the formula

$$°C = 0.56(°F - 32)$$

Here is an example of how to use the formula. We know that water boils at 212°F (at sea level). To find the boiling point of water in degrees Celsius, we proceed as follows:

$$°C = 0.56(°F - 32)$$
$$°C = 0.56(212 - 32)$$
$$°C = 0.56(180)$$
$$°C = 100$$

Thus, in a few simple steps, we find that the boiling point of water in the metric system is 100°C.

_____ **3** Using the formula and following the example given in Step 2 above, find the following temperatures in °C:

a. Water freezes at sea level at 32°F.

32°F = _______°C

b. A mixture of 50 percent water and 50 percent antifreeze in the radiator will protect the engine against freezing to −34°F.

−34°F = _______°C

c. The temperature indicator light on a new car is set to come on at 260°F.

260°F = _______°C

_____ **4** Most of us have always measured pressure in pounds per square inch (psi). However, many specifications for pressure are now given in the metric engineering unit, kilograms per square centimeter (kg/cm^2). To convert a pressure from psi to kg/cm^2 you use the formula

$$kg/cm^2 = psi \times 0.07$$

For example, the oil pressure of an engine is given as 40 psi. To find the pressure in kg/cm^2, proceed as follows:

$$kg/cm^2 = psi \times 0.07$$
$$kg/cm^2 = 40 \times 0.07$$
$$kg/cm^2 = 2.8$$

_____ **5** Using the formula and following the example given in Step 4 above, find the following pressure in kg/cm^2: The fuel pump pressure on a car is given as 3 to 5 psi.

3 to 5 psi = _______ to _______ kg/cm^2

Written Checkup — Job 32

In the spaces below, write down the general specifications for the engine in your car. (These specifications can be found in a shop auto repair manual.) Then use the metric conversion tables in the Appendix to find each of the specifications in the metric system. When you are finished, have your instructor check your work.

GENERAL ENGINE SPECIFICATIONS

Horsepower	_______ hp	_______ kW
Bore	_______ in	_______ mm
Stroke	_______ in	_______ mm
Compression ratio	_______	_______
Displacement	_______ in^3	_______ cm^3
Oil pressure	_______ psi	_______ kg/cm^2

Metric Conversion Tables

MILLIMETERS TO INCHES

mm	1	2	3	4	5	6	7	8	9	10	11	12	13
Inches	0.0394	0.0787	0.1181	0.1575	0.1968	0.2362	0.2756	0.3150	0.3543	0.3937	0.4331	0.4724	0.5118
mm	14	15	16	17	18	19	20	21	22	23	24	25	26
Inches	0.5512	0.5905	0.6299	0.6693	0.7087	0.7480	0.7874	0.8268	0.8661	0.9055	0.9449	0.9842	1.0236
mm	27	28	29	30	31	32	33	34	35	36	37	38	39
Inches	1.0630	1.1024	1.1417	1.1811	1.2205	1.2598	1.2992	1.3386	1.3779	1.4173	1.4567	1.4961	1.5354
mm	40	41	42	43	44	45	46	47	48	49	50	51	52
Inches	1.5748	1.6142	1.6535	1.6929	1.7323	1.7716	1.8110	1.8504	1.8898	1.9291	1.9685	2.0079	2.0472
mm	53	54	55	56	57	58	59	60	61	62	63	64	65
Inches	2.0866	2.1260	2.1653	2.2047	2.2441	2.2835	2.3228	2.3622	2.4016	2.4409	2.4803	2.5197	2.5590
mm	66	67	68	69	70	71	72	73	74	75	76	77	78
Inches	2.6984	2.6378	2.6772	2.7165	2.7559	2.7953	2.8346	2.8740	2.9134	2.9527	2.9921	3.0315	3.0709
mm	79	80	81	82	83	84	85	86	87	88	89	90	91
Inches	3.1102	3.1496	3.1890	3.2283	3.2677	3.3071	3.3464	3.3858	3.4252	3.4646	3.5039	3.5433	3.5827
mm	92	93	94	95	96	97	98	99	100				
Inches	3.6220	3.6614	3.7008	3.7401	3.7795	3.8189	3.8583	3.8976	3.9370				

INCHES TO MILLIMETERS

Inches	1/64	1/32	3/64	1/16	5/64	3/32	7/64	1/8	9/64	5/32	11/64	3/16	13/64
mm	0.3969	0.7937	1.1906	1.5875	1.9844	2.3812	2.7781	3.1750	3.5719	3.9687	4.3656	4.7625	5.1594
Inches	7/32	15/64	1/4	17/64	9/32	19/64	5/16	21/64	11/32	23/64	3/8	25/64	13/32
mm	5.5562	5.9531	6.3500	6.7469	7.1437	7.5406	7.9375	8.3344	8.7312	9.1281	9.5250	9.9219	10.3187
Inches	27/64	7/16	29/64	15/32	31/64	1/2	33/64	17/32	35/64	9/16	37/64	19/32	39/64
mm	10.7156	11.1125	11.5094	11.9062	12.3031	12.7000	13.0969	13.4937	13.8906	14.2875	14.6844	15.0812	15.4781
Inches	5/8	41/64	21/32	43/64	11/16	45/64	23/32	47/64	3/4	49/64	25/32	51/64	13/16
mm	15.8750	16.2719	16.6687	17.0656	17.4625	17.8594	18.2562	18.6531	19.0500	19.4469	19.8437	20.2406	20.6375
Inches	53/64	27/32	55/64	7/8	57/64	29/32	59/64	15/16	61/64	31/32	63/64		
mm	21.0344	21.4312	21.8281	22.2250	22.6219	23.0187	23.4156	23.8125	24.2094	24.6062	25.0031		

FAHRENHEIT TO CELSIUS (Centigrade)

°F	−20	−15	−10	−5	0	1	2	3	4	5	10	15	20
°C	−28.9	−26.1	−23.3	−20.6	−17.8	−17.2	−16.7	−16.1	−15.6	−15.0	−12.2	−9.4	−6.7
°F	25	30	35	40	45	50	55	60	65	70	75	80	85
°C	−3.9	−1.1	1.7	4.4	7.2	10.0	12.8	15.6	18.3	21.1	23.9	26.7	29.4
°F	90	95	100	105	110	115	120	125	130	135	140	145	150
°C	32.2	35.0	37.8	40.6	43.3	46.1	48.9	51.7	54.4	57.2	60.0	62.8	65.6
°F	155	160	165	170	175	180	185	190	195	200	205	210	212
°C	68.3	71.1	73.9	76.7	79.4	82.2	85.0	87.8	90.6	93.8	96.1	98.9	100.0
°F	215	220	225	230	235	240	245	250	255	260	265		
°C	101.7	104.4	107.2	110.0	112.8	115.6	118.3	121.1	123.9	126.6	129.4		

FEET TO METERS

ft	0	1	2	3	4	5	6	7	8	9	ft
	m	m	m	m	m	m	m	m	m	m	
–		0.305	0.610	0.914	1.219	1.524	1.829	2.134	2.438	2.743	–
10	3.048	3.353	3.658	3.962	4.267	4.572	4.877	5.182	5.486	5.791	10
20	6.096	6.401	6.706	7.010	7.315	7.620	7.925	8.230	8.534	8.839	20
30	9.144	9.449	9.754	10.058	10.363	10.668	10.973	11.278	11.582	11.887	30
40	12.192	12.497	12.802	13.106	13.411	13.716	14.021	14.326	14.630	14.935	40
50	15.240	15.545	15.850	16.154	16.459	16.764	17.069	17.374	17.678	17.983	50
60	18.288	18.593	18.898	19.202	19.507	19.812	20.117	20.422	20.726	21.031	60
70	21.336	21.641	21.946	22.250	22.555	22.860	23.165	23.470	23.774	24.079	70
80	24.384	24.689	24.994	25.298	25.603	25.908	26.213	26.518	26.822	27.127	80
90	27.432	27.737	28.042	28.346	28.651	28.956	29.261	29.566	29.870	30.175	90
100	30.480	30.785	31.090	31.394	31.699	32.004	32.309	32.614	32.918	33.223	100

Source: Buick Motor Division of General Motors Corporation

MILES TO KILOMETERS

mile	0	1	2	3	4	5	6	7	8	9	mile
	km	km	km	km	km	km	km	km	km	km	
—		1.609	3.219	4.828	6.437	8.047	9.656	11.265	12.875	14.484	—
10	16.093	17.703	19.312	20.921	22.531	24.140	25.750	27.359	28.968	30.578	10
20	32.187	33.796	35.406	37.015	38.624	40.234	41.843	43.452	45.062	46.671	20
30	48.280	49.890	51.499	53.108	54.718	56.327	57.936	59.546	61.155	62.764	30
40	64.374	65.983	67.593	69.202	70.811	72.421	74.030	75.639	77.249	78.858	40
50	80.467	82.077	83.686	85.295	86.905	88.514	90.123	91.733	93.342	94.951	50
60	96.561	98.170	99.779	101.39	103.00	104.61	106.22	107.83	109.44	111.04	60
70	112.65	114.26	115.87	117.48	119.09	120.70	122.31	123.92	125.53	127.14	70
80	128.75	130.36	131.97	133.58	135.19	136.79	138.40	140.01	141.62	143.23	80
90	144.84	146.45	148.06	149.67	151.28	152.89	154.50	156.11	157.72	159.33	90
100	160.93	162.54	164.15	165.76	167.37	168.98	170.59	172.20	173.81	175.42	100

SQUARE INCHES TO SQUARE CENTIMETERS

in^2	0	1	2	3	4	5	6	7	8	9	in^2
	cm^2	cm^2	cm^2	cm^2	cm^2	cm^2	cm^2	cm^2	cm^2	cm^2	
—		6.452	12.903	19.355	25.806	32.258	38.710	45.161	51.613	58.064	—
10	64.516	70.968	77.419	83.871	90.322	96.774	103.226	109.677	116.129	122.580	10
20	129.032	135.484	141.935	148.387	154.838	161.290	167.742	174.193	180.645	187.096	20
30	193.548	200.000	206.451	212.903	219.354	225.806	232.258	238.709	245.161	251.612	30
40	258.064	264.516	270.967	277.419	283.870	290.322	296.774	303.225	309.677	316.128	40
50	322.580	329.032	335.483	341.935	348.386	354.838	361.290	367.741	374.193	380.644	50
60	387.096	393.548	399.999	406.451	412.902	419.354	425.806	432.257	438.709	445.160	60
70	451.612	458.064	464.515	470.967	477.418	483.870	490.322	496.773	503.225	509.676	70
80	516.128	522.580	529.031	535.483	541.934	548.386	554.838	561.289	567.741	574.192	80
90	580.644	587.096	593.547	599.999	606.450	612.902	619.354	625.805	632.257	638.708	90
100	645.160	651.612	658.063	664.515	670.966	677.418	683.870	690.321	696.773	703.224	100

CUBIC INCHES TO CUBIC CENTIMETERS

in^3	0	1	2	3	4	5	6	7	8	9	in^3
	cm^3	cm^3	cm^3	cm^3	cm^3	cm^3	cm^3	cm^3	cm^3	cm^3	
—		16.387	32.774	49.161	65.548	81.935	98.322	114.709	131.097	147.484	—
10	163.871	180.258	196.645	213.032	229.419	245.806	262.193	278.580	294.967	311.354	10
20	327.741	344.128	360.515	376.902	393.290	409.677	426.064	442.451	458.838	475.225	20
30	491.612	507.999	524.386	540.773	557.160	573.547	589.934	606.321	622.708	639.095	30
40	655.483	671.870	688.257	704.644	721.031	737.418	753.805	770.192	786.579	802.966	40
50	819.353	835.740	852.127	868.514	884.901	901.289	917.676	934.063	950.450	966.837	50
60	983.224	999.611	1015.998	1032.385	1048.772	1065.159	1081.546	1097.933	1114.320	1130.707	60
70	1147.094	1163.482	1179.869	1196.256	1212.643	1229.030	1245.417	1261.804	1278.191	1294.578	70
80	1310.965	1327.352	1343.739	1360.126	1376.513	1392.200	1409.288	1425.675	1442.062	1458.449	80
90	1474.836	1491.223	1507.610	1523.997	1540.384	1556.771	1573.158	1589.545	1605.932	1622.319	90
100	1638.706	1655.093	1671.481	1687.863	1704.255	1720.642	1737.029	1753.416	1769.803	1786.190	100

CUBIC FEET TO CUBIC METERS

ft^3	0	1	2	3	4	5	6	7	8	9	ft^3
	m^3	m^3	m^3	m^3	m^3	m^3	m^3	m^3	m^3	m^3	
—		0.0283	0.0566	0.0850	0.1133	0.1416	0.1699	0.1982	0.2265	0.2549	—
10	0.2832	0.3115	0.3398	0.3681	0.3964	0.4248	0.4531	0.4814	0.5097	0.5380	10
20	0.5663	0.5947	0.6230	0.6513	0.6796	0.7079	0.7362	0.7646	0.7929	0.8212	20
30	0.8495	0.8778	0.9061	0.9345	0.9628	0.9911	1.0194	1.0477	1.0760	1.1044	30
40	1.1327	1.1610	1.1893	1.2176	1.2459	1.2743	1.3026	1.3309	1.3592	1.3875	40
50	1.4159	1.4442	1.4725	1.5008	1.5291	1.5574	1.5858	1.6141	1.6424	1.6707	50
60	1.6990	1.7273	1.7557	1.7840	1.8123	1.8406	1.8689	1.8972	1.9256	1.9539	60
70	1.9822	2.0105	2.0388	2.0671	2.0955	2.1238	2.1521	2.1804	2.2087	2.2370	70
80	2.2654	2.2937	2.3220	2.3503	2.3786	2.4069	2.4353	2.4636	2.4919	2.5202	80
90	2.5485	2.5768	2.6052	2.6335	2.6618	2.6901	2.7184	2.7468	2.7751	2.8034	90
100	2.8317	2.6800	2.8884	2.9167	2.9450	2.9733	3.0016	3.0300	3.0583	3.0866	100

GALLONS (U.S.) TO LITERS

U.S. gal	0	1	2	3	4	5	6	7	8	9	U.S. gal
	l	l	l	l	l	l	l	l	l	l	
—		3.7854	7.5709	11.3563	15.1417	18.9271	22.7126	26.4980	30.2834	34.0638	—
10	37.8543	41.6397	45.4251	49.2105	52.9960	56.7814	60.5668	64.3523	68.1377	71.9231	10
20	75.7085	79.4940	83.2794	87.0648	90.8502	94.6357	98.4211	102.2065	105.9920	109.7774	20
30	113.5528	117.3482	121.1337	124.9191	128.7045	132.4899	136.2754	140.0608	143.8462	147.6316	30
40	151.4171	155.2025	158.9879	162.7734	166.5588	170.3442	174.1296	177.9151	181.7005	185.4859	40
50	189.2713	193.0568	196.8422	200.6276	204.4131	208.1985	211.9839	215.7693	219.5548	223.3402	50
60	227.1256	230.9110	234.6965	238.4819	242.2673	246.0527	249.8382	253.6236	257.4090	261.1945	60
70	264.9799	268.7653	272.5507	276.3362	280.1216	283.9070	287.6924	291.4779	295.2633	299.0487	70
80	302.8342	306.6196	310.4050	314.1904	317.9759	321.7613	325.5467	329.3321	333.1176	336.9030	80
90	340.6884	344.4738	348.2593	352.0447	355.8301	359.6156	363.4010	367.1864	370.9718	374.7573	90
100	378.5427	382.3281	386.1135	389.8990	393.6844	397.4698	401.2553	405.0407	408.8261	412.6115	100

POUNDS TO KILOGRAMS

lb	0	1	2	3	4	5	6	7	8	9	lb
	kg	kg	kg	kg	kg	kg	kg	kg	kg	kg	
—		0.454	0.907	1.361	1.814	2.268	2.722	3.175	3.629	4.082	—
10	4.536	4.990	5.443	5.897	6.350	6.804	7.257	7.711	8.165	8.618	10
20	9.072	9.525	9.979	10.433	10.886	11.340	11.793	12.247	12.701	13.154	20
30	13.608	14.061	14.515	14.969	15.422	15.876	16.329	16.783	17.237	17.690	30
40	18.144	18.597	19.051	19.504	19.958	20.412	20.865	21.319	21.772	22.226	40
50	22.680	23.133	23.587	24.040	24.494	24.948	25.401	25.855	26.308	26.762	50
60	27.216	27.669	28.123	28.576	29.030	29.484	29.937	30.391	30.844	31.298	60
70	31.751	32.205	32.659	33.112	33.566	34.019	34.473	34.927	35.380	35.834	70
80	36.287	36.741	37.195	37.648	38.102	38.555	39.009	39.463	39.916	40.370	80
90	40.823	41.277	41.730	42.184	42.638	43.092	43.545	43.998	44.453	44.906	90
100	45.359	45.813	46.266	46.720	47.174	47.627	48.081	48.534	48.988	49.442	100

POUNDS PER SQUARE INCH TO KILOGRAMS PER SQUARE CENTIMETER

lb/in^2	0	1	2	3	4	5	6	7	8	9	lb/in^2
	kg/cm^2	kg/cm^2	kg/cm^2	kg/cm^2	kg/cm^2	kg/cm^2	kg/cm^2	kg/cm^2	kg/cm^2	kg/cm^2	
—		0.0703	0.1406	0.2100	0.2812	0.3515	0.4218	0.4921	0.5625	0.6328	—
10	0.7031	0.7734	0.8437	0.9140	0.9843	1.0546	1.1249	1.1952	1.2655	1.3358	10
20	1.4062	1.4765	1.5468	1.6171	1.6874	1.7577	1.8280	1.8983	1.9686	2.0389	20
30	2.1092	2.1795	2.2498	2.3202	2.3905	2.4608	2.5311	2.6014	2.6717	2.7420	30
40	2.8123	2.8826	2.9529	3.0232	3.0935	3.1639	3.2342	3.3045	3.3748	3.4451	40
50	3.5154	3.5857	3.6560	3.7263	3.7966	3.8669	3.9372	4.0072	4.0779	4.1482	50
60	4.2185	5.2888	4.3591	4.4294	4.4997	4.5700	4.6403	4.7106	4.7809	4.8512	60
70	4.9216	4.9919	5.0622	5.1325	5.2028	5.2731	5.3434	5.4137	5.4840	5.5543	70
80	5.6246	5.6949	5.7652	5.8356	5.9059	5.9762	6.0465	6.1168	6.1871	6.2574	80
90	6.3277	6.3980	6.4683	6.5386	6.6089	6.6793	6.7496	6.8199	6.8902	6.9605	90
100	7.0308	7.1011	7.1714	7.2417	7.3120	7.3823	7.4526	7.5229	7.5933	7.6636	100

POUND-FEET TO KILOGRAM-METERS

lb-ft	0	1	2	3	4	5	6	7	8	9	lb-ft
	kg-m	kg-m	kg-m	kg-m	kg-m	kg-m	kg-m	kg-m	kg-m	kg-m	
—		0.138	0.276	0.415	0.553	0.691	0.829	0.967	1.106	1.244	—
10	1.382	1.520	1.658	1.796	1.934	2.073	2.211	2.349	2.487	2.625	10
20	2.764	2.902	3.040	3.178	3.316	3.455	3.593	3.731	3.869	4.007	20
30	4.146	4.284	4.422	4.560	4.698	4.837	4.975	5.113	5.251	5.389	30
40	5.528	5.666	5.804	5.942	6.080	6.219	6.357	6.495	6.633	6.771	40
50	6.910	7.048	7.186	7.324	7.462	7.601	7.739	7.877	8.015	8.153	50
60	8.292	8.430	8.568	8.706	8.844	8.983	9.121	9.259	9.397	9.535	60
70	9.674	9.812	9.950	10.088	10.227	10.365	10.503	10.641	10.779	10.918	70
80	11.056	11.194	11.332	11.470	11.609	11.747	11.885	12.023	12.161	12.300	80
90	12.438	12.576	12.714	12.855	12.991	13.129	13.267	13.405	13.544	13.682	90
100	13.820	13.958	14.096	14.235	14.373	14.511	14.649	14.787	14.925	14.064	100

Answer Key for *Automotive Tools, Fasteners, and Measurements*, by William H. Crouse and Donald L. Anglin

Answers for Multiple-Choice Tests

Correct Answer	Workbook Job Number
B	3
A	27
D	17
A	31
C	1
A	—
C	6
C	9
B	18
D	—
D	20
C	11
A	28
B	7
A	
D	
B	
C	
B	
D	
C	
B	
A	

Answers for True-or-False Tests

Question	Workbook Job Number																		
	2	**4**	**5**	**8**	**10**	**12**	**13**	**14**	**15**	**16**	**19**	**21**	**22**	**23**	**24**	**25**	**26**	**29**	**30**
1	T	T	F	F	T	T	T	F	T	F	F	F	F	T	F	T	F	F	T
2	F	F	T	T	T	F	T	T	F	F	T	F	T	F	T	T	T	T	T
3	T	F	F	T	F	T	F	T	T	T	T	T	F	T	T	F	T	T	F
4	T	T	T	T	T	T	F	F	T	F	T	F	T	T	F	F	F	T	T
5	T	F	T	T	F	T	T	T	F	T	T	T	T	T	F	T	F	T	T
6	F	T	T		T	T					F	T	T	T				F	F
7	T	T	T		T	F					F	T	T	F				F	F
8	T	F	F			F					T	T	F	F				T	T
9	F	F	F			T					F	F	F	T				F	F
10	T	T	T			T					T	F	T	T				T	T

